PROFESSIONAL BURNOUT

Series in Applied Psychology: Social Issues and Questions

Stevan E. Hobfoll, *Editor-in-Chief*

PROFESSIONAL BURNOUT: RECENT DEVELOPMENTS IN THEORY AND RESEARCH

Edited by

Wilmar B. Schaufeli
University of Nijmegen, The Netherlands

Christina Maslach
University of California, Berkeley, USA

Tadeusz Marek
Jagiellonian University, Kraków, Poland

Taylor & Francis

USA	Publishing Office:	Taylor & Francis
		1101 Vermont Avenue, N.W., Suite 200
		Washington, DC 20005-3521
		Tel: (202) 289-2174
		Fax: (202) 289-3665
	Distribution Center:	Taylor & Francis
		1900 Frost Road, Suite 101
		Bristol, PA 19007-1598
		Tel: (215) 785-5800
		Fax: (215) 785-5515
UK		Taylor & Francis Ltd.
		4 John Street
		London WC1N 2ET, UK
		Tel: 071 405 2237
		Fax: 071 831 2035

PROFESSIONAL BURNOUT: Recent Developments in Theory and Research

1 2 3 4 5 6 7 8 9 0   B R B R   9 8 7 6 5 4 3

This book was set in Times Roman by Taylor & Francis. The editors were Janice Stern and Deena Williams Newman; the production supervisor was Peggy M. Rote; and the typesetters were Anahid Alvandian and Darrell Larsen. Cover design by Michelle Fleitz.
Printing and binding by Braun-Brumfield, Inc.

A CIP catalog record for this book is available from the British Library.

⊗ The paper in this publication meets the requirements of the ANSI standard
  Z39.48-1984(Permanence of Paper).

Library of Congress Cataloging-in-Publication Data

Professional burnout: recent developments in theory and research /
  edited by Wilmar B. Schaufeli, Christina Maslach, Tadeusz Marek.
    p.   cm. — (Series in applied psychology)
    Includes bibliographical references and index.

    1. Burnout (Psychology)   2. Job stress.   3. Stress (Psychology)
  I. Schaufeli, Wilmar,   date.   II. Maslach, Christina.
  III. Marek, Tadeusz.   IV. Series: Series in applied psychology (New
  York, N.Y.)
    [DNLM: 1. Burnout, Professional—congresses. WM 172 P9635]
BF481.P77   1993
158.7—dc20
DNLM/DLC
for Library of Congress   92-49131
ISBN 1-56032-262-4   CIP
ISSN 1048-8146

CONTENTS

III   ORGANIZATIONAL APPROACHES

CONTRIBUTORS

ROBERT A. BOUDREAU, Faculty of Management, University of Lethbridge, Lethbridge, Alberta, Canada T1K 3N4.

MATTHIAS BURISCH, Department of Psychology, University of Hamburg, von Melle Park 5, 2000 Hamburg 13, Germany.

BRAM P. BUUNK, Department of Psychology, University of Groningen, Grote Kruistraat 1/2, 9712 TS Groningen, The Netherlands.

CARY CHERNISS, Graduate School of Applied and Professional Psychology, Rutgers University, P.O. Box 819, New Brunswick, New Jersey 08855, USA.

TOM COX, Centre for Organizational Health and Development, Department of Psychology, University of Nottingham, Nottingham NG7 2RD, United Kingdom.

DIRK ENZMANN, Department of Psychology, Technical University, Dovestrasse 1-5, 1000 Berlin 10, Germany.

MAGDALENA FAFROWICZ, Institute of Psychology, Jagiellonian University, Golebia 13, 31-007 Kraków, Poland.

ROBERT T. GOLEMBIEWSKI, Department of Political Science, University of Georgia, Athens, Georgia 30602, USA.

JOHN FREEDY, Medical University of South Carolina, 171 Ashley Avenue, Charleston, South Carolina 29425, USA.

NOËLLE GIRAULT, Department of Psychology, Réné Descartes University, 28 Rue Serpente, 75006 Paris, France.

LENNART HALLSTEN, National Institute of Occupational Health, Division of Psychophysiology, Ekelundsvägen 16, 171 84 Solna, Sweden.

STEVAN E. HOBFOLL, Applied Psychology Center, Kent State University, Kent, Ohio 44242, USA.

GEORGE KUK, Centre for Organizational Health and Development, Department of Psychology, University of Nottingham, Nottingham NG7 2RD, United Kingdom.

MICHAEL P. LEITER, Department of Psychology, Acadia University, Wolfville, and Dalhousie University, Halifax, Nova Scotia, Canada B0P 1XO.

TADEUSZ MAREK, Institute of Psychology, Jagiellonian University, Golebia 13, 31-007 Kraków, Poland.

CHRISTINA MASLACH, Department of Psychology, University of California, Berkeley, California 94720, USA.

CZESLAW NOWOROL, Institute of Psychology, Jagiellonian University, Golebia 13, 31-007 Kraków, Poland.

AYALA M. PINES, Department of Psychology, University of California, Berkeley, California 94720, USA.

WILMAR B. SCHAUFELI, Department of Psychology, University of Nijmegen, P.O. Box 9104, 6500 HE Nijmegen, The Netherlands.

KATHERINE SCHERB, Department of Political Science, University of Georgia, Athens, Georgia 30602, USA.

JACQUES WINNUBST, Centre for Psychology of Health and Illness, University of Utrecht, Heidelberglaan 1, 3508 TC Utrecht, The Netherlands.

ZBIGNIEW ŽARCZYŃSKI, Institute of Psychology, Jagiellonian University, Golebia 13, 31-007 Kraków, Poland.

PREFACE

Our goal for this book was a challenging one: to give a complete presentation of the past, present, and future of burnout. To achieve this goal, we have brought together a set of original papers from an international group of leading scholars on burnout.

The inspiration for this book was the first European Conference on Professional Burnout, which was held in Kraków, Poland, in 1990. This conference was organized by the European Network of Organizational Psychologists (ENOP), the Unit of Work and Organizational Psychology of the University of Nijmegen (The Netherlands), and the Department of Work Psychology and Ergonomics of Jagiellonian University, Kraków. Financial support was provided by the Center de l'Étude de Sciences de l'Homme (Paris, France) and by the Polish Ministry of Education (Governmental Research Grant No. III.29). The conference included all the major burnout theorists and researchers from the United States, Canada, The Netherlands, Germany, Sweden, Great Britain, France, Israel, and Poland. Many of these scholars were meeting each other for the first time, although they were familiar with each other's work. The small size and the interactive format of the conference promoted a more exciting exchange of ideas than is typical of most professional meetings, and several critical issues were raised and debated. In many ways, this conference represented a key turning point in the historical development of the burnout concept, in terms of both resolving some past questions and opening up some new ones. Clearly, the significance of the entire event called for a permanent record of its achievements. Thus this book was born.

This volume is not, however, merely a varied collection of convention papers. Instead, the chapters were commissioned by the editors from specific participants, based on developments that had taken place during the conference itself. In some cases the chapters are completely new contributions that pull together some of the ideas that emerged at the meeting and/or provide major reviews of key areas. In other cases the chapters build on and extend the original conference presentation, and a few have been revised for a more specific focus. Our editorial goal in shaping these chapters was to produce a state-of-the-art analysis of burnout that would be both comprehensive and integrative. We believe we have been successful in achieving our aim, and we hope that this volume will challenge and inspire future generations of burnout scholars.

Nijmegen/San Francisco/Kraków
May 1992

The Editors

1

HISTORICAL AND CONCEPTUAL DEVELOPMENT OF BURNOUT

Christina Maslach
University of California, Berkeley

Wilmar B. Schaufeli
University of Nijmegen

"It is a critical time for the concept of burnout. Will burnout prove to be a concept of enduring value, useful in understanding and treating a class of work-related symptoms? Or will the concept itself 'burn out' from overuse, overextension, and lack of new direction?" (Farber, 1983b, pp. 17–18). It has been a decade since this "critical time" and in the intervening years there has been an extraordinary amount of work on this topic. In their recent bibliography, Kleiber and Enzmann (1990) listed nearly 1,500 publications that were published in the 7 years following Farber's 1983 statement. This is even more than in the previous decade 1974–1983, when 1,000 books, journal articles, and dissertations on burnout were written. Although the early work on burnout was almost exclusively American in origin, the current contributions are truly international in scope.

What have we learned from this extensive empirical and clinical literature? Is burnout really a concept of enduring value that improves our understanding of the working world of many professionals? We believe that the answer is yes—an assessment that is well supported by the evidence documented in this volume. However, before beginning our tour through that territory, we first need to explore its definitional borders. In other words, how do we conceptualize, on various dimensions, the nature of burnout? To answer this question, we need to review the developmental history of the burnout concept and to trace the pattern of theoretical progress that has been made.

HISTORICAL DEVELOPMENT OF THE BURNOUT CONCEPT

Burnout first emerged as a social problem, not as a scholarly construct. Thus, the initial conception of burnout was shaped by pragmatic rather than academic concerns. In this pioneering phase of conceptual development, the focus was on clinical descriptions of burnout. Later on, there was a second, empirical phase, in which the emphasis shifted to systematic research on burnout and in particular to the assessment of this phenomenon. Throughout these two phases there has been increasing theoretical development in which the concern has been to integrate the evolving notion of burnout with other conceptual frameworks.

The Pioneer Phase

The first few articles about burnout appeared in the mid-1970s in the United States (Freudenberger, 1974, 1975; Maslach, 1976). The significance of these first articles was that they provided an initial description of the burnout phenomenon, gave it its name, and showed that it was not an aberrant response by a few deviant people but was actually more common.

The way in which the burnout phenomenon was identified and labeled illustrates its social origin. As a psychiatrist, Freudenberger was employed in an alternative health care agency. He observed that many of the volunteers with whom he was working experienced a gradual emotional depletion and a loss of motivation and commitment. Generally, this process took about a year and was accompanied by a variety of mental and physical symptoms. To denote this particular mental state of exhaustion, Freudenberger used a word that was being used colloquially to refer to the effects of chronic drug abuse: "burnout."

At about the same time, Maslach, a social psychology researcher, was studying the ways in which people cope with emotional arousal on the job (see Chapter 2). She was particularly interested in such cognitive strategies as "detached concern" and "dehumanization in self-defense," but soon discovered that both the arousal and the strategies had important implications for people's professional identity and job behavior. When by chance she described these results to an attorney, she was told that poverty lawyers called this particular phenomenon "burnout." Once Maslach and her colleagues adopted this term, they discovered that it was immediately recognized by their interviewees; thus, a new colloquial expression was born.

Do these anecdotes about the "discovery" of burnout indicate that the phenomenon did not exist before? Obviously not. For instance, Burisch (Chapter 5) presents several examples of psychological states that have been described previously in the literature. These states match the current description of burnout but have been labeled differently (e.g., "exhaustion reaction"). In 1953, Schwartz and Will published a case study of Miss Jones, a disillusioned psychiatric nurse, who is probably the most prominent (and often cited) example of burnout. Moreover, essayists have portrayed fictional burned-out characters long before the concept was introduced in the mid-1970s. For instance, Thomas Mann's description of the protagonist in *Buddenbrooks* (1922) includes the most essential features of burnout, such as extreme fatigue and the loss of idealism and passion for one's job. Most famous, however, is Graham Greene's *A Burnt Out Case* (1960), in which a spiritually tormented and disillusioned architect quits his job and withdraws into the African jungle. The symptoms displayed by this character fit quite well with current descriptions of burnout.

Given these forerunners, why did the burnout syndrome not attract public attention until the mid-1970s? Several authors point to a specific constellation of economic, social, and historical factors. Farber suggests that "American workers have become increasingly disconnected and alienated from their communities, and increasingly insistent upon attaining personal fulfillment and gratification from their work. The combination of these two trends has produced workers with higher expectations of fulfillment and fewer recourses to cope with frustrations—a perfect recipe for burnout" (1983a, p. 11). Farber also points to a problematic development in the human services. Originally these services were based in the community, but after World War II social services work became more professionalized, bureaucratized, credentialized, and isolated. Governmental interference increased, and clients became needier and more entitled to services. Consequently, it became more difficult for people to find professional fulfillment in human services work, and disillusionment and burnout became increasingly common.

In addition, Cherniss (1980a) argued that the tendency toward individualization in modern society has led to increasing pressure on the human services. Because the traditional social fabric is disintegrating, more and more problems in living have to be solved by professionals instead of relatives, neighbors, or other members of the community. Furthermore, over the past decades, the government has cut back costs for many human service agencies, so that an increasing workload has to be managed by even fewer people. Cherniss (1980a) also points to the decline of the authority of professionals over the past decades and to the recent development of what he calls the "professional mystique." The latter involves the public's belief that professionals experience a high level of autonomy and job satisfaction, are highly trained and competent, work with responsive clients, and are generally compassionate and caring. This mystique is reinforced by the professionals' education and leads to high and unrealistic expectations in young professionals that clash with the harsh everyday reality of the job.

Thus, the stage was set for the introduction of the burnout concept in the mid-1970s. There was a tremendous reaction to the first articles on this topic, and in the next 5 years there was a virtual flood of writing about burnout. Obviously, burnout was "in the air." Not surprisingly, interest was particularly high among practitioners, as opposed to academic scholars, and thus much of the writing on burnout appeared in magazines or journals directed to this professional audience. These practitioners came from a variety of people-oriented, human services occupations where (1) the relationship between a provider and a recipient is central to the job and (2) the provision of service, care, or education can be fraught with emotional strain. Consequently, the greatest attention to burnout, and the most discussion of it in those first 5 years, occurred in the fields of education, social services, medicine, the criminal justice system, mental health, religion, and various other people-oriented occupations (see Maslach, 1982a for an annotated review).

This early burnout literature had several noteworthy characteristics, which in turn have had implications for the development of the burnout concept. First, what was meant by the term "burnout" varied widely from one writer to the next. As a result, these writers were sometimes talking about different phenomena rather than the same one. A second, and related, characteristic is that the concept of burnout was stretched and expanded to encompass far more than it did originally. Almost every personal problem that one can think of was described as "burnout" at some point. In some cases, burnout was a somewhat superfluous addition, as in "midlife crisis burnout." In other cases, it was stretched to include opposite phenomena, such as overload and underload. The problem here is that a concept that has been expanded to mean everything ends up meaning nothing at all, an issue that has been discussed elsewhere (Maslach, 1982c; Maslach & Jackson, 1984).

A third characteristic of the early burnout literature is that it was largely nonempirical. Perlman and Hartman (1982) reviewed 48 articles that had been published between 1974 and 1981, all with many ideas, suggestions, and proposals about what causes burnout and what should be done about it. However, only five of these articles (i.e., 10%) had any empirical data beyond an occasional anecdote or personal case history. Instead, most of these early articles on burnout used a "clinical" approach. Thus, the authors tried to describe and understand the burnout syndrome by means of cautious (but unstandardized) observations and subsequent analyses of individual case studies. In particular, the focus was on symptoms that are found in burned-out individuals.

Most early articles on burnout followed a typical pattern. First, the stressful nature of the particular profession was described. Next, job stress in that profession was related to burnout, and one or more case studies or vignettes were presented to illustrate the issue. Finally, some preventive strategies were recommended. Thus, although much was being said about burnout, there was little evidence to either support or refute these statements. This relative lack of empirical evidence limited any attempts at intervention and led critics to disparage the concept or even dismiss it entirely.

One reason for the initial sparseness of research on burnout is that, as mentioned earlier, practitioners were far more interested in burnout than were academic scholars. This is not surprising, given that practitioners are more likely to be dealing directly with the problem of burnout on a daily basis. However, this fact has had some important ramifications. Most practitioners have had less training in research skills and thus are less likely to undertake major research projects. Furthermore, practitioners have different goals with respect to burnout than do academicians. Their primary concern is with intervention rather than theory, i.e., with how to solve the problem rather than with how to conceptualize it. Most academic scholars would argue that that is putting the cart before the horse and that one must first have a theoretical model about the phenomenon in order to know what to do about it. However, people who are actually dealing with the problem tend to view theory building as intellectual game playing, with no practical payoff. The argument is that we already know enough about burnout from direct experience, so now we should apply that knowledge to alleviate the problem.

On the other hand, many researchers were not interested in burnout at first, even though (or maybe because) practitioners were. Initially, the academic world had a somewhat negative reaction to the concept of burnout. "Because it has a catchy ring to it, burnout is sometimes immediately dismissed as a fad or as pseudoscientific jargon that is all surface flash and no substance" (Maslach & Jackson, 1984a, p. 139). For instance, Maslach and Jackson's psychometric article on the development of the Maslach Burnout Inventory (MBI) was returned by some journal editors with a short note that it had not even been read "because we do not publish 'pop' psychology."

Because there was not an early emphasis on developing theories of burnout, there was no conceptual framework for integrating and evaluating the various findings and proposed solutions. Consequently, the field seemed somewhat scattered and chaotic. However, this lack of an initial theoretical model was not as serious a flaw as some have argued. To some extent, the lack of theory reflected the newness of the phenomenon—much had to be discovered about its parameters before a model could be developed. Also, a different process occurs when one starts with a real-world problem and works back toward a theoretical model rather than vice versa (starting with a theory and then deriving some implications for a particular social issue). Real problems tend to be messy and complex rather than clean and simple, and no single theory is going to be the obvious choice as a guiding model. Moreover, different people will work back toward different theoretical models for the same problem, depending on their particular perspective. For example, someone with a clinical perspective may conceptualize burnout in terms of depression, but someone with an organizational perspective may approach it as an issue of job satisfaction. Initially, it can be hard to compare and integrate these different perspectives (especially if there is definitional variation as well), and this may be one reason why the earlier burnout literature lacked theoretical coherence. However, as has been pointed out elsewhere, this rich diversity of theoretical perspectives, as well as of related methodological techniques, is one of the special virtues of the eclectic, problem-oriented approach that has typified the burnout field (Maslach & Jackson, 1984).

The Empirical Phase

During the next phase of the 1980s, the work on burnout entered a more focused, constructive, and empirical period. Many books and articles were written about burnout, in which the authors outlined their working models of the phenomenon, proposed various ideas and interventions, and presented various forms of corroborative evidence (survey and questionnaire data, interview responses, clinical case studies). Standardized measures of burnout were developed, thus providing researchers with more precise definitions and methodological tools for studying the phenomenon (for a review, see Chapter 12). In particular, the development and widespread acceptance of the Maslach Burnout Inventory (MBI; Maslach & Jackson, 1981a, 1981b, 1986) and the Tedium Measure (TM; Pines, Aronson, & Kafry, 1981) fostered systematic research on burnout, resulting in an increased number of articles published in scholarly journals (including several issues devoted entirely to burnout).

Up until the early 1980s burnout was studied exclusively in the United States. Gradually, the phenomenon drew attention in other countries as well, beginning with such English-speaking countries as Canada and Great Britain. Soon articles and books were being translated into other languages, and by the second half of the 1980s, research instruments (particularly the MBI) were also being translated into French, German, Italian, Spanish, Swedish, Dutch, Polish, and Hebrew, among others (see also Chapter 12). Soon, the first cross-national studies on burnout were carried out (for a review see Chapter 13).

Interestingly, in other countries burnout research started after the concept had been established in the United States, and after measurement instruments had been developed. Accordingly, what we have called the pioneer phase in the development of burnout was skipped in countries outside the United States. In these countries, burnout was conceptualized, from the very beginning, in operational terms that were implied by the measurement instruments. Thus, for researchers using the MBI, burnout was by definition a syndrome characterized by emotional exhaustion, depersonalization, and reduced personal accomplishment. Hence, the initial conceptual debate on burnout was less broad, and alternative measures were rarely developed. Only recently have conceptual contributions been made by non–Anglo-Saxon authors (see Chapters 4–6, 9, and 10).

A general review of the more recent burnout literature indicates several trends. First, much of the work has continued to be done within people-oriented, human service occupations, although the variety of these occupations has expanded (e.g., police, correctional officers, prison guards, librarians). Although this is the general case, the burnout concept has also been extended to other types of occupations and other nonoccupational areas of life. For example, there has been discussion about burnout in the business world, in sports, in political activism, and within the family.

The empirical research on burnout has tended to focus more on job factors than on other types of variables. This is consistent with most of the conceptual models that have been proposed. Thus, researchers have studied such variables as job satisfaction, job stress (workload, role conflict, and role ambiguity), job withdrawal (turnover, absenteeism), job expectations, relations with coworkers and supervisors (social support on the job), relations with clients, caseload, type of position and time in job, agency policy, and so forth. The personal factors that have been studied are most often demographic variables (sex, age, marital status, etc.). In addition, some attention has been given to personality variables (locus of control, hardiness), personal health, relations with family and friends (social support at home), and personal values and commitment. In general, job factors are more strongly related to burnout than are biographical or personal factors.

The vast majority of the empirical work on burnout consists of correlational studies that collect subjective, self-report data (mostly employing the MBI or the TM) at one point in time from a nonrepresentative sample. Although some interesting findings have come from this research, it is important to recognize its limitations. First, some of the correlations between burnout and different variables may be an artifact of the reliance on a single method (common method variance) or the use of a specialized group (selection effects). Second, response rates tend to be rather low, which could indicate that particularly burned out respondents do not fill out the questionnaires because they feel embarrassed or threatened (another selection effect). Third, such studies do not permit a test of causal hypotheses, even though causal links are usually presumed and discussed. Fourth, the subjective assessments of certain variables may not accord with their objective status (e.g., a supervisor might provide helpful advice, but an employee might perceive it as condescending and thus report a lack of supervisor support). This is an interesting research question in its own right, but it is mentioned here only in terms of its methodological implications.

Given these limitations, one must be cautious in interpreting empirical data. For example, as indicated above, higher burnout has been correlated with poor job conditions of various kinds, and a common conclusion is that these job conditions have caused people to burn out. However, it may be that people who are experiencing burnout begin to see everything in a negative light and report that the job conditions are poor (whether they are or not). Both interpretations are interesting in terms of our basic understanding of burnout, but their conceptual implications, as well as their implications for intervention, are very different.

Recently, several longitudinal studies of burnout were conducted (Dignam & West, 1988; Firth & Britton, 1989; Golembiewski & Munzenrider, 1988; Jackson, Schwab, & Schuler, 1986; Shirom, 1986; Wade, Cooley, & Savicki, 1986; Wolpin, 1988). These more methodologically sophisticated studies lead to three major conclusions. First, the level of burnout seems fairly stable over time. Obviously, its nature is more chronic than acute. Second, burnout leads to physical symptoms, to absenteeism, and to job turnover. Third, role conflict and lack of social support from colleagues and supervisors are antecedents of burnout.

Unfortunately, many of the studies, including the ones with longitudinal designs, have not been grounded in a theoretical framework. That is, they have not utilized a conceptual model of burnout, from which hypotheses are derived and tested. In many cases, there has not even been a clear rationale for the choice of variables. This atheoretical stance can cause further problems in interpretation of the results. It makes it difficult to assess whether the findings are supportive of the researcher's ideas about burnout, whether the findings are due to chance, whether the findings are consistent with other relevant research, and whether there is a clear interpretation of a lack of significant results (incorrect hypothesis? methodological problem? chance?).

However, in recent years much progress has been made on the theoretical front. One factor that has helped facilitate this progress is the greater consensual agreement on an operational definition of burnout, largely because of the development of validated research measures. Good opportunities now exist for integrating empirical results within a particular conceptual framework and for carrying out theory-driven research. At this stage of theoretical development there are various conceptual models from which researchers can choose to guide their empirical studies. These models are presented in the first three parts of this volume. Some of these models are current reformulations of earlier conceptual frameworks whereas some are new models that are being articulated for the first time. Thus, this volume contains the most comprehensive presentation of current theories of burnout, which we hope will both enrich ongoing work in the field and inspire further theoretical development.

CONCEPTUAL ISSUES

Although much progress has been made, and there is the promise of more, some issues pertaining to the specificity of burnout still need to be addressed. In particular, there are three important questions. First, is burnout a distinctive syndrome that can be distinguished from other related concepts, such as job stress, depression, or job dissatisfaction? Second, is the experience of burnout limited to human services professions, or is it a more general phenomenon that is also found in other occupations or even outside the work sphere? Third, are there diagnostic criteria that would allow burnout to be identified within an individual?

Is Burnout a Distinctive Concept?

How is burnout distinguished from other psychological constructs? Is burnout truly a new phenomenon or is it simply an "old" phenomenon with a new label? The diversity of burnout causes, symptoms, definitions, and consequences has contributed much to the confusion about the specificity of burnout. For instance, at various times burnout has been equated with tedium, (job) stress, (job) dissatisfaction, (reactive or professional) depression, alienation, low morale, anxiety, (job) strain, tension, feeling "worn out," experiencing "flame-out," tension, conflict, pressure, "nerves," boredom, (chronic or emotional) fatigue, poor mental health, crisis, helplessness, vital exhaustion, and hopelessness. As Cox, Kuk, and Leiter (Chapter 11) argue, part of this confusion is caused by mixing two different levels of explanation: the level of common discourse and the level of scientific definition and measurement. This result is probably due to the fact that burnout emerged as a pragmatic issue rather than developing within a scholarly context (as previously mentioned).

In this section we will concentrate on the distinctiveness of burnout from other related and established psychological constructs. In particular, questions have been raised about the extent to which burnout can be distinguished from such concepts as stress (especially job stress), depression, and job dissatisfaction. It should be noted that most of these other concepts are plagued with the same sort of definitional ambiguity as burnout. Thus, the problem of specifying burnout is by no means an exception.

Burnout can only be distinguished in a relative way from other related concepts. There are no sharp boundaries, and trying to establish such divisions could be very artificial. However, a relative distinction between burnout and "stress" can be made with respect to time, and between burnout and both "depression" and "satisfaction" with respect to domain.

Burnout can be considered as prolonged job stress, i.e., demands at the workplace that tax or exceed an individual's resources. This longer time perspective is also implied in the terminology: burning out (depleting one's resources) is a long-term process. A remarkable parallel exists with the work of Selye (1967), the "founding father" of the concept of stress. According to Selye, exposure to a stressor leads to the general adaptation syndrome, consisting of three phases: alarm, resistance, and exhaustion. In the final phase, after prolonged exposure to stress, the physiological resources are depleted, and irreversible damage is caused to the organism. Referring to Selye's adaptation syndrome, Etzion (1987) argued that burnout is a latent process of psychological erosion resulting from prolonged exposure to job stress. Typically, the exhaustion phase is reached before the individual consciously has noticed both preliminary phases. Brill (1984) has also conceptualized burnout as prolonged job stress. According to Brill, stress refers to an adaptation process that is temporary and is accompanied by mental and physical symptoms, whereas burnout refers to a breakdown in adaptation accompanied by chronic malfunctioning.

Therefore, the relative distinction with respect to time between (job) stress and burnout implies that both concepts can only be discriminated retrospectively when the adaptation has been successfully performed (job stress) or when a breakdown in adaptation has occurred (burnout). To put it in another way, stress and burnout cannot be distinguished on the basis of their symptoms, but only on the basis of the process. Regarded from this viewpoint, it is remarkable that burnout has been studied predominantly as a state and not as a process that develops over time. This criticism is emphasized by Burisch (Chapter 5), Hallsten (Chapter 6), and Leiter (Chapter 14), each of whom proposes a different approach to the study of burnout *process*.

A fundamental problem arises when one tries to make a distinction with respect to the domain between burnout and other concepts. Various theoretical perspectives predict that burnout should be related to such concepts as depression and job dissatisfaction. But at what point does a relationship become so strong that both concepts are reinterpreted as indices of the same underlying construct? In other words, "being different from" and "being related to" are not mutually exclusive. This distinction becomes even harder to make when measurement and methodological problems are taken into account. For instance, a high correlation between two concepts can be due to several artifacts (e.g., common method variance), whereas a poor correlation might be caused by unidentified confounding variables. Nevertheless, it is a legitimate question to ask whether burnout can be differentiated, conceptually as well as empirically, from such affective states as depression and job dissatisfaction.

It has been claimed by Freudenberger (1981) that (reactive) depression is most often accompanied by guilt, whereas burnout generally occurs in the context of anger. Unfortunately, only clinical evidence is presented for this assertion. Moreover, Freudenberger argues that the symptoms of burnout, at least initially, tend to be job-related and situation-specific rather than pervasive. A "real" depression is characterized by a generalization of the person's symptoms across all situations. In a somewhat similar vein, Warr (1987) distinguished between two types of affective well-being: depression is considered to be "context-free," whereas burnout is regarded as "job-related." Oswin (1978) described a syndrome of "professional depression" among nurses that bears close resemblance to burnout, including being overtired, becoming hardened, and accepting one's ineffectuality at the job. Thus, on a conceptual level there seems to be some agreement about the specificity of burnout as a job-related syndrome, which is characterized by dysphoric symptoms that are similar to those of depression.

Empirically speaking, research on the discriminant validity of the MBI shows that the emotional exhaustion component of burnout is substantively related to depression. The relationships with depersonalization and personal accomplishment are less strong (for a review, see Chapter 12). Hence, the conclusion of some authors that burnout and depression show a considerable overlap (e.g., Meier, 1984) is only partly correct. The fact that depression is differently related to the three MBI dimensions underscores the validity of a multidimensional model of burnout (see Chapter 2).

A similar argument can be made with respect to the relationship between job satisfaction and the three burnout dimensions. Most empirical studies with the MBI show that job satisfaction is negatively correlated with emotional exhaustion and depersonalization, but (contrary to expectations) only weakly correlated with personal accomplishment. In most studies only correlations are reported (for a review, see Chapter 12). However, a German study by Enzmann and Kleiber (1989) uses a more comprehensive factor analysis of scale means and suggests that personal accomplishment and job satisfaction constitute one factor.

The overall pattern of the research findings has led some researchers to conclude that while burnout and job dissatisfaction are clearly linked, they are not identical constructs (Zedeck, Maslach, Mosier, & Skitka, 1988). However, the specific nature of that link is still a matter of speculation. One possibility is that burnout causes a drop in job satisfaction, i.e., that job dissatisfaction is an affective outcome of burnout. The reverse causal hypothesis is that job dissatisfaction causes burnout. Alternatively, both burnout and job dissatisfaction may be caused by a third variable, such as poor working conditions.

Given the current state of knowledge on these issues, the most conclusive statement that we can make is that the distinctiveness of the burnout concept pertains to both its process (time) and to its multidimensionality (domain). Hence, what is needed in future work on burnout is (1) a greater emphasis on process research and (2) further development of multifactorial models. Several of the chapters in this volume can serve as starting points for such endeavors. We return to this issue in Chapter 15.

Is Burnout Limited to the Human Services?

Much of the burnout research has continued to be done within people-oriented, human service occupations, although the variety of these occupations has expanded, as seen previously. The syndrome was first and foremost visible in the human services, perhaps because the negative stereotyping of clients dramatically runs counter to what is considered to be a professional attitude. The concept of burnout is still very relevant for these groups.

However, burnout is not necessarily restricted to the human services professions. Because burnout appears to be specific to the work domain, in the sense that its origins lie in the job situation, the phenomenon might be found in other types of occupations. For example, there has been discussion about burnout in business, and corporate or managerial burnout (e.g., Cahoon & Rowney, 1984; Etzion, Kafry, & Pines, 1982; Ginsburg, 1974; Levinson, 1981). Burnout has also appeared in the sports world, in terms of both coaches (e.g., Caccese & Mayerberg, 1984; Capel, Sisley, & Desertrain, 1987) and athletes (e.g., Fender, 1989; Smith, 1986). In addition, the burnout construct has been extended to nonjob domains. For example, burnout is being studied among voluntary activities, such as political activism (Gomes & Maslach, 1991; Pines, in press-b). Furthermore, burnout has been applied to the family sphere, as evidenced by analyses of parent burnout (Pelsma, Roland, Tollefson, & Wigington, 1989; Procaccini & Kiefaber, 1983) and marriage burnout (Pines, 1988).

Although the concept of burnout has been extended beyond the traditional borders of the human services, there has been no analysis of the form that burnout may take in these different professions. Thus, an issue that arises is whether the burnout definition (i.e., the MBI) has been transplanted uncritically to those fields and is being used in ways that are not as meaningful or relevant. For example, in occupations that do not involve work with clients, what is the meaning of the personal accomplishment and the depersonalization components of burnout? In some studies (e.g., Golembiewski & Munzenrider, 1988), the MBI has been modified to refer to relationships with coworkers rather than relationships with clients. What has been preserved is the interaction with people, but what has changed is whether these particular people are the core element of the job.

According to some critics, this change transforms the basic concept rather than simply translating it. For instance, referring to the rewording of MBI items, Garden writes:

> Whether "getting along well with co-workers" in a research and development organization is a salient dimension for feelings of personal accomplishment is highly questionable. It would seem more likely that solving technical problems, or getting a report out on time would be the salient variable leading to feelings of personal accomplishment. Similarly, whether "caring about what happens to co-workers" for a research and development worker is comparable to "caring about what happens to clients" for the human services worker is doubtful. The meaning of such items with the word change is highly ambiguous. (1987a, p. 548)

The latter assertion is supported empirically by Schaufeli and Peeters (1990). They found that the internal consistencies of the depersonalization and personal accomplishment scales were unsatisfactory in non–human-services samples, which employed an adapted version of the MBI (in which the term "recipients" had been replaced by "coworkers").

Accordingly, what is needed is a thorough analysis of the core elements of particular jobs that would allow researchers to adapt the three burnout dimensions to other occupations. Such a reanalysis might involve a change in the labels for these dimensions. For example, the core element of an executive job is the responsibility for the continuity of the organization. Hence, executive burnout might be characterized by a negative attitude toward this responsibility (e.g., "the profit I make just flows into the pockets of the shareholders"). The core element for a research and development worker is solving technical problems under time pressure. Accordingly, burnout in this profession might be manifested by a negative attitude toward technology (e.g., "by improving modern technology, I am helping to destroy nature"). This type of reconceptualization should focus on types of professions (rather than single jobs), and the design of any revised measurement instruments should begin (as was the case with the MBI) with extensive interviews of workers in these professions.

The extension of the burnout concept to nonoccupational domains has been more controversial. According to some critiques, the term burnout takes on a different meaning when it is being applied to marriage or parenting, as opposed to a job in the human services. If burnout is being used in these contexts as simply a synonym for unhappiness, or frustration, or dissatisfaction with one's spouse or child, then it is a superfluous addition that is not adding any new insights to these issues. However, given that the basic conceptual framework of burnout is based on the social relationship between provider and recipient, it could be argued that such a concept could be applied in meaningful ways to other types of relationships, such as parent–child or husband–wife (see Chapter 2). Such an application would require the appropriate translation of the various burnout dimensions into these family contexts.

Are There Individual Diagnostic Criteria for Burnout?

To answer this question, we must first take a closer look at various definitions of burnout in which the process of burnout is emphasized: for example, "Burnout is a process that begins with excessive and prolonged levels of job tension. This stress produces strain in the worker (feelings of tension, irritability, and fatigue). The process is completed when the workers defensively cope with the job stress by psychologically detaching themselves from the job and becoming apathetic, cynical, and rigid" (Cherniss, 1980b, p. 40). Similarly, burnout has been characterized as "a progressive loss of idealism, energy and purpose experienced by people in the helping professions as a result of the conditions of their work" (Edelwich & Brodsky, 1980, p. 14). The burnout process has also been described as follows: "to deplete oneself; to exhaust one's physical and mental resources; to wear oneself out by excessively striving to reach some unrealistic expectation imposed by oneself or by the values of society" (Freudenberger & Richelson, 1980, p. 16).

This gradual process of "burning out" (see Chapter 6) can result in a particular state that has been identified as burnout by some authors. Paine (1982a, pp. 6–7) distinguished between the burnout stress syndrome ("the identifiable cluster of feelings and behaviors most commonly found in stressful or highly frustrating work environments") and burnout mental disability ("the often serious, clinically significant pattern of personal distress and diminished performance which is an end stage of the burnout process"). According to Paine, the burnout stress syndrome is not typically a mental disorder, but it gradually develops over time and may eventually result in mental disability.

Diagnostic criteria can only be applied to this end state of the process, which has been defined in several ways. Freudenberger and Richelson (1980) describe a number of clinical symptoms of burnout, including exhaustion, detachment, boredom and cynicism, impatience and heightened irritability, a sense of omnipotence, a suspicion of being unappreciated, paranoia, disorientation, denial of feelings, and psychosomatic complaints. Pines and Aronson (1988) define burnout as a state of physical, emotional, and mental exhaustion caused by a long-term involvement in situations that are emotionally demanding, and describe this state as follows: "Physical exhaustion is characterized by low energy, chronic fatigue, and weakness" (p. 11); "Emotional exhaustion, the second component of burnout, involves primarily feelings of helplessness, hopelessness, and entrapment" (p. 13); "Mental exhaustion, the third component, is characterized by the development of negative attitudes towards one's self, work, and life itself" (p. 13).

The most widely used definition of burnout comes from Maslach and Jackson (1986, p. 1): "Burnout is a syndrome of emotional exhaustion, depersonalization and reduced personal accomplishment that can occur among individuals who do 'people work' of some kind." They go on to say that "burnout can lead to a deterioration in the quality of care or service provided . . . it appears to be a factor in job turnover, absenteeism, and low morale . . . it seems to be correlated with various self-reported indices of personal dysfunction, including physical exhaustion, insomnia, increased use of alcohol and drugs, and marital and family problems" (1986, p. 2). Finally, a less well-known but rather precise "operational definition of burnout" was presented by Brill: "an exceptionally mediated, job-related, dysphoric and dysfunctional state in an individual without major psychopathology who has (1) functioned for a time at adequate performance and affective levels in the same job situation and who (2) will not recover to previous levels without outside help or environmental rearrangement" (1984, p. 14).

Although these "state" definitions of burnout differ in scope and precision, they share at least five common elements. First, there is a predominance of dysphoric symptoms such as mental or emotional exhaustion, fatigue, and depression. Second, the accent is on mental and behavioral symptoms rather than on physical symptoms, although some authors mention atypical physical complaints as well. Third, burnout symptoms are work-related. Fourth, the symptoms manifest themselves in "normal" persons who did not suffer from psychopathology before. Fifth, decreased effectiveness and work performance occur because of negative attitudes and behaviors.

Based on a similar analysis of definitions, Bibeau et al. (1989) propose subjective and objective diagnostic criteria for burnout. The principal subjective indicator is a general state of severe fatigue accompanied by (1) loss of self-esteem resulting from a feeling of professional incompetence and job dissatisfaction; (2) multiple physical symptoms of distress without an identifiable organic illness; and (3) problems in concentration, irritability, and negativism. The principal objective indicator of burnout is a significant decrease in work performance over a period of several months, which has to be observable in relation to (1) recipients (who receive services of lesser quality); (2) supervisors (who observe a decreasing effectiveness, absenteeism, etc.); and (3) colleagues (who observe a general loss of interest in work-related issues). Bibeau et al. (1989) also mention three criteria of exclusion that allow a differential diagnosis. These subjective and objective indicators of burnout should not result from (1) sheer incompetence (i.e., the person has to have performed well in the job for a significant period), (2) major psychopathology, or (3) family-related problems. Also, severe fatigue resulting from monotonous work or a big workload is excluded because this is not necessarily accompanied by feelings of incompetence or lowered productivity.

Despite these concrete diagnostic criteria, Bibeau et al. (1989) conclude that it would be superfluous to introduce a new psychiatric nosographic category such as burnout or professional exhaustion. In their opinion, such a mental state is included in the subcategory of "adjustment disorders with work (or academic) inhibition" of the DSM-III (American Psychiatric Association, 1980), which currently is the most widely employed diagnostic tool in mental health. According to DSM-III (p. 299), an adjustment disorder is characterized by "a maladaptive reaction to an identifiable psycho-social stressor, that occurs within three months after the onset of the stressor. The maladaptive nature of the reaction is indicated by either impairment in social or occupational functioning or symptoms that are in excess of a normal and expected reaction to the stressor." More specifically, the adjustment disorder with work (or academic) inhibition should be used when "the predominant manifestation is an inhibition in work or academic functioning occurring in an individual whose previous work or academic performance has been adequate. Frequently there are also varying mixtures of anxiety and depression" (p. 301).

Whether Bibeau et al. (1989) are correct in arguing that a new diagnostic category is unnecessary, they have demonstrated that burnout can be assessed in psychiatric terms. This means that standardized questionnaires, particularly the MBI, have the potential to be used in the individual assessment of burnout. The MBI covers the major aspects of burnout and has good psychometric properties (see Chapter 12). Nevertheless, clinically validated cutoff points have to be developed in order to apply it for individual assessment.

However, the description of burnout in psychiatric terms (as defined in the DSM-III) may have the negative consequence of labeling individuals as mentally ill. It has been argued that one reason for the popularity of the burnout concept is that it is a socially accepted label that carries a minimal stigma (Shirom, 1989). From this perspective, the introduction of a psychiatric diagnosis of burnout would be a step backward. On the other hand, in most countries such a diagnosis is necessary to determine if a worker is entitled to a leave of absence, treatment, or other benefits. Thus, the benefit of a psychiatric diagnosis for burnout would be that it would provide an official recognition of a legitimate personal problem. In considering the pros and cons of this issue, we should keep in mind that such a diagnosis would refer to the end stage of a long process; presumably, relatively few people would reach this state.

CONCLUSION

Now that we have reviewed the past history and conceptual development of burnout, it is time to turn to a discussion of its current status, both theoretical and empirical. Presented in the following pages are a wealth of ideas and critiques by the leading scholars in the burnout field. Of course, this book is not the final, definitive statement on burnout; many questions and controversies remain, and much theorizing and research have yet to be done. But this volume makes many significant contributions to the ongoing discussion. First, it presents a variety of theoretical perspectives that link burnout to concepts that have been developed in social psychology (Part I), personality and clinical psychology (Part II), and organizational psychology (Part III). Second, it presents some state-of-the-art assessments of crucial research issues in the study of burnout (Part IV). By integrating the latest developments in theory and research into this volume, we hope that our book will have a major impact on our future understanding of professional burnout.

I

INTERPERSONAL APPROACHES

Traditionally, the major cause of burnout has been the emotionally demanding interpersonal relationships of professional caregivers with their recipients. By definition, these relationships are asymmetric. Professionals in the human services provide care, support, attention, comfort, and assistance to their clients, patients, and pupils. The unbalanced interpersonal relationship is nicely illustrated semantically by the terms "caregiver" and "recipient." The former gives, the latter receives. Eventually, the strains of this asymmetric relationship may result in the depletion of the caregiver's emotional resources, the core symptom of burnout.

In the first contribution, Maslach advocates a complex multifaceted model of burnout. She describes the emergence of the three-dimensional burnout concept, which is derived empirically rather than theoretically. The first component—emotional exhaustion—closely resembles an orthodox stress variable. Both other components add the crucial social dimension to burnout. Depersonalization refers to the person's negative perception of his or her recipients, whereas reduced personal accomplishment includes a person's negative self-evaluation in relation to his or her job performance. Essentially, Maslach argues that burnout is an individual stress experience that originates from emotionally demanding interpersonal relationships with recipients. This experience is embedded in a complex social context and also involves the person's attitude toward both others and self.

Buunk and Schaufeli agree with Maslach's three-dimensional model of burnout. They show empirically that each dimension is characterized by a different set of correlates, which suggests that distinct psychological processes are involved. Even more important, they extend the social context in which burnout develops beyond the emotionally demanding relationship with recipients. In their view, not only do social exchange processes with recipients play a role, but social comparison and social affiliation processes with colleagues and superiors are relevant etiological factors for burnout. For instance, Buunk and Schaufeli found some evidence for a process similar to emotional contagion, in which persons with a high need for social comparison adopt burnout symptoms that they perceive in their colleagues. Moreover, the authors demonstrated the relevance of personality characteristics, such as self-esteem, for understanding social affiliation and comparison processes at work.

According to Pines, burnout results from a failure in the existential quest for meaning. The people who are in danger of burning out are those who are highly motivated and strongly devoted to their jobs. In other words, people who expect to derive a sense of existential significance from their work are likely candidates for burnout. Although her existential approach to burnout is somewhat general, Pines agrees that burnout occurs most often in professionals who work with people. They experience burnout because of their long term involvement in emotionally demanding interpersonal interactions with recipients. Moreover, in the human services the "dedicatory ethic" is widespread, which means that work is considered a calling rather than a job. The resulting risk is that human services professionals tend to look for existential meaning in their relationships with recipients. Pines illustrates the relevance of her existential perspective for the treatment of burnout by discussing two exercises, one of which ("your best and worst clients") highlights the central importance of interpersonal relationships.

The contributions to this part of the book demonstrate that interpersonal factors are essential in understanding burnout. In important ways, burnout is linked to emotionally demanding relationships with recipients (Maslach) and also to relationships with coworkers and supervisors (Buunk & Schaufeli). Moreover, burnout is reinforced by a failure to find existential meaning in these interpersonal relationships, particularly with recipients in the human services (Pines).

2

BURNOUT: A MULTIDIMENSIONAL PERSPECTIVE

Christina Maslach
University of California, Berkeley

It has been almost 20 years since the term *burnout* first appeared in the psychological literature. The phenomenon that was portrayed in those early articles had not been entirely unknown, but had been rarely acknowledged or even openly discussed. In some occupations it was almost a taboo topic because it was considered tantamount to admitting that at times professionals can (and do) act "unprofessionally." The reaction of many people was to deny that such a phenomenon existed, or, if it did exist, to attribute it to a very small (but clearly mentally disturbed) minority. This response made it difficult, at first, to take any work on burnout seriously. However, after the initial articles were published, there was a major shift in opinion. Professionals in the human services gave substantial support to both the validity of the phenomenon and its significance as an occupational hazard. Once burnout was acknowledged as a legitimate issue, it began to attract the attention of various researchers.

Our knowledge and understanding of burnout have grown dramatically since that shaky beginning. Burnout is now recognized as an important social and individual problem. There has been much discussion and debate about the phenomenon, its causes, and its consequences. As these ideas about burnout have proliferated, so have the number of empirical research studies to test these ideas. We can now begin to speak of a body of work about burnout, much of which is reviewed and cited in the current volume. This work is now viewed as a legitimate and worthy enterprise that has the potential to yield both scholarly gains and practical solutions. What I would like to do in this chapter is give a personal perspective on the concept of burnout.

Having been one of the "pioneers" in this field, I have the advantage of a long-term viewpoint that covers the 20 years from the birth of the concept of burnout to its present proliferation. Furthermore, because my research was among the earliest, it has had an impact on the development of the field. In particular, my definition of burnout and my measure to assess it (Maslach Burnout Inventory, or MBI) have been adopted by many researchers and have thus influenced subsequent theorizing and research. My work has also been the point of departure for various critiques. Thus, for better or for worse, my perspective on burnout has played a part in framing the field, and so it seems appropriate to articulate that viewpoint. In presenting this perspective, however, I do not intend to simply give a summary statement of ideas that I have discussed elsewhere. Rather, I want to provide a retrospective review and analysis of why those ideas developed in the ways that they did. Looking back on my work, with the hindsight of 20 years, I can see more clearly how my research path was shaped by both choice and chance. The shape of that path has had some impact on which questions have been asked about burnout (and which have not) as well as on the manner in which answers have been sought. A better understanding of the characteristics of that path will, I think, provide some insights into our current state of knowledge and debate about burnout.

In some sense, this retrospective review marks a return to my research roots. The reexamination of my initial thinking about burnout, and an analysis of how that has developed and changed over the years, has led me to renew my focus on the core concept of social relationships. I find it appropriately symbolic that this return to my research roots occurred within the context of a return to my ancestral roots. The 1990 burnout conference that inspired this rethinking took place in southern Poland, from which each of my paternal grandparents, Michael Maslach and Anna Pszczolkowska, emigrated to the United States in the early 1900s. Thus, my travel to Krakow had great significance for me at both personal and professional levels.

A MULTIDIMENSIONAL MODEL OF BURNOUT

The operational definition, and the corresponding measure, that is most widely used in burnout research is the three-component model developed by Susan Jackson and myself (Maslach & Jackson, 1981a, 1981b, 1984a, 1986). We define burnout as a psychological syndrome of emotional exhaustion, depersonalization, and reduced personal accomplishment that can occur among individuals who work with other people in some capacity. Emotional exhaustion refers to feelings of being emotionally overextended and depleted of one's emotional resources. Depersonalization refers to a negative, callous, or excessively detached response to other people, who are usually the recipients of one's service or care. Reduced personal accomplishment refers to a decline in one's feelings of competence and successful achievement in one's work.

This definition of burnout did not derive from an existing theory but was developed on the basis of several years of exploratory research. This research involved interviews, surveys, and field observations of employees in a wide variety of "people-oriented" professions, including health care, social services, mental health, criminal justice, and education.

Early Research

Unlike my research on other topics, my work on burnout did not begin with a clearly defined phenomenon or a particular theoretical model. Indeed, the research did not even begin with a focus on burnout at all. Instead, my interest was in emotion and in the general question of how people "know" what they are feeling. I had been doing some experimental research on how people interpret an ambiguous arousal state—what cues they use and what information they seek to make sense of this uncertain feeling (Maslach, 1979). My approach to this issue was a social cognitive one. That is, I was concerned with the cognitive processes by which people attached a label to an experience and thus gave it meaning. One of the intriguing theoretical possibilities was that the same arousal state could be interpreted as very different emotions, depending on the cognitive label (an argument articulated by Schachter and Singer, 1962).

As I continued to work in this area, I became interested in a somewhat different question, which was inspired by the then-ongoing research on misattribution processes (Ross, Rodin, & Zimbardo, 1969; Storms & Nisbett, 1970). Suppose people experienced a strong arousal state (rather than an ambiguous one). Could similar cognitive processes be used to relabel this arousal and thus change its meaning and its experiential impact? More specifically, was such cognitive relabeling a means of reducing the intensity of the arousal and thus preventing it from disrupting necessary, ongoing behavior? Such a question would be especially important for people who have to function calmly and efficiently in situations that are often characterized by crisis and chaos. For example, rescue personnel (such as police or firefighters) have to deal with such situations, as do staff of hospital emergency rooms and therapists doing crisis counseling.

My initial review of psychological theories and constructs found little that addressed this issue directly. However, there were two concepts from the medical literature that seemed relevant. One of these was "detached concern" (Lief & Fox, 1963), which referred to the medical profession's ideal of blending compassion with emotional distance. Although the practitioner is concerned about the patient's well-being, he or she recognizes that it is necessary to avoid overinvolvement with the patient and to maintain a detached objectivity.

The second relevant concept was "dehumanization in self-defense" (Zimbardo, 1970), which referred to the process of protecting oneself from overwhelming emotional feelings by responding to other people more as objects than as persons. For example, if a patient has a condition that is upsetting to see or otherwise difficult to work with, it may be easier for the practitioner to provide the necessary care if he or she thinks of the patient as a particular "case" or "symptom" rather than as a human being who is suffering.

Both of these concepts seemed to shed some theoretical light on the issue of how people cope with strong emotional arousal. However, there were many questions about how medical personnel used these concepts in real-life situations. For example, are detachment and dehumanization purely cognitive constructs, or are they expressed in overt behaviors that could affect the relationship between practitioner and patient? Are these techniques used selectively, in response to certain situations only (and if so, what are the critical factors in these situations), or are they used continuously as part of a "professional manner"? Is there any explicit training in the use of such techniques? Does their use produce only temporary effects, in terms of the reduction of strong arousal, or are there more long-term consequences? It seemed to me that these (and many other) questions were deserving of some serious study.

Case Study Interviews

Given that the two guiding concepts had their origins in the medical professions, my first step was to interview people who were working in health care settings. Thus, the initial, exploratory interviews were with physicians and nurses. Subsequent interviews were conducted with people working in the area of mental health, including psychiatrists, psychiatric nurses, and hospice counselors.

Some key themes emerged from these interviews. First, it became clear that emotional experiences played an important role in the provision of health care. Some of these experiences were enormously rewarding and uplifting, as when patients recovered because of the practitioner's efforts. However, other experiences were emotionally stressful for the practitioner, such as working with difficult or unpleasant patients, having to give "bad news" to patients or their families, dealing with patient deaths, or having conflicts with coworkers or supervisors. Such emotional strains were sometimes overwhelming, and practitioners talked about being emotionally exhausted and drained of all feeling.

A second theme was that "detached concern" was often more an impossible ideal than an attainable reality. Although practitioners would try to distance and detach themselves from sources of emotional strain, they found it difficult to do so and still maintain concern. A more typical pattern was a negative shift over time, in terms of practitioners' perceptions and feelings about their patients; in extreme cases, they began to dislike and even despise them. It was this phenomenon that gradually became the topic of interest rather than the original notion of detached concern.

A third general theme had to do with the self-assessment of professional competence. All too often the experience of emotional turmoil was interpreted as a failure to "be professional" (i.e., nonemotional, cool, objective) and led people to question their ability to work in a health career. Many practitioners felt that their formal training had not prepared them for the emotional reality of their work and its subsequent impact on their personal functioning.

Up to this point, my thinking about this phenomenon had been framed within the context of health care. However, my focus was broadened as the result of a chance event. I happened to describe the results of my health interviews to an attorney, who told me that a similar phenomenon, called "burnout," occurred among poverty lawyers working in legal services. Not only did I learn that the phenomenon I was studying had a name, but I learned that it was present in a wider range of occupations. What seemed to link poverty law and health care was the focus on providing aid and service to people in need—in other words, the core of the job was the relationship between provider and recipient. The implication was that working with other people, particularly in a caregiving relationship, was at the heart of the burnout phenomenon.

To assess the validity of this idea, I began a second phase of exploratory interviews with providers in other types of "people work" occupations (which included poverty lawyers, ministers, teachers, prison guards, and probation officers). The decision to use each of these subject populations was somewhat serendipitous, in that informants in one group would often refer us to people in the next. Similar themes emerged from these interviews, although the specific content differed as a function of the type of occupation (e.g., differences between working with students or delinquent adolescents). This evidence of a parallel pattern suggested that burnout was not just some idiosyncratic response to stress but was a syndrome with some identifiable regularities (Maslach, 1976).

It was very clear from all of these interviews that the provision of service or care can be a very demanding and involving occupation, and that emotional exhaustion is not an uncommon response to such job overload. Freudenberger's report of both his own and others' experiences as staff members in "alternative" institutions provided similar evidence. Indeed, Freudenberger placed particular emphasis on the centrality of exhaustion for burnout, describing it as an end state of exhaustion caused by excessive demands on one's energy and resources (Freudenberger, 1974, 1975). It should be noted that the notion of exhaustion presupposes a prior state of high arousal or overload rather than one of low arousal or underload. Thus, this definitional component of emotional exhaustion stands in contrast to some other conceptualizations, which view burnout as a response to tedious, boring, and monotonous work.

The second component of depersonalization also emerged from these interviews as human services employees described how they tried to cope with the emotional stresses of their work. Moderating one's compassion for clients by emotional distance from them (detached concern) was viewed as a way of protecting oneself from intense emotional arousal that could interfere with effective functioning on the job. However, an imbalance of excessive detachment and little concern seemed to lead staff to respond to clients in negative, callous, and dehumanized ways (Maslach, 1973). Thus, excessive detachment, or depersonalization, could impair performance and be detrimental to the quality of care.

Interwoven throughout the interviews was a central focus on relationships—usually between provider and recipient, but also between the provider and coworkers or family members. These relationships were the source of both emotional strains and rewards, and sometimes they functioned as a resource for coping with stress. The centrality of these interactions for the experiences that were being described suggested that an interpersonal analysis of the overall phenomenon would be most appropriate.

Questionnaire Surveys

At this stage in my research I thought it necessary to change my methodological approach. The interviews had generated many ideas based on the individual case studies, but now it was time to investigate these ideas using other sources of data. First of all, I wanted to study burnout among larger samples rather than relying on the reports of a few people. Second, I wanted to develop more systematic assessment techniques so that I could both compare and aggregate the responses of these larger groups. Third, I wanted to study burnout within its situational context rather than simply looking at the stress experience of the individual provider in isolation. With these goals in mind, I embarked on a series of questionnaire survey studies.

The initial pilot study (conducted with Kathy Kelly) involved the social work staff in a major urban welfare agency. All of the staff completed questionnaires, and some of them also participated in interviews. In addition, observations were made on site. The results of this pilot study were very fruitful. We were able to draft a questionnaire in which our interview questions were transformed into standardized response formats and then to refine this questionnaire based on respondent feedback. We also began to get a better feel for the situational context of the provider–recipient relationship as a consequence of our on-site observations. Thus, we could see first-hand some of the job factors that had been described in earlier interviews, such as the high number of clients (caseload), prevalence of negative client feedback, and scarcity of resources. Moreover, we were able to observe other aspects of the interaction between social worker and client, such as nonverbal "distancing" behaviors, which seemed relevant to burnout but had not been reported by previous providers (indeed, the providers may not have been aware of them). In particular, we were able to observe the behaviors of the clients as well as of the providers. Consequently, the focus was on both sides of the helping relationship (as opposed to just the perspective of the provider), which led me to develop some ideas about the role of the client in the burnout process (Maslach, 1978a).

Once the pilot study was successfully completed, the next step was to conduct some systematic survey research on burnout. The first two studies (done in collaboration with Ayala Pines) were designed to assess providers' emotional states and reactions to their clients (the two burnout dimensions of emotional exhaustion and depersonalization) and to discover if these dimensions were correlated with certain job factors. The basic questionnaire that had been developed in the pilot study with social workers was amended to be appropriate for two different subject populations: staff of day care centers and staff of several types of mental health institutions. In the mental health study (Pines & Maslach, 1978), the focus continued to be on the role of the recipient (in this case, patients) in burnout. The day care study (Maslach & Pines, 1977) also investigated caseload issues, but added some organizational factors (program structure, staff participation in decision making). Upon learning about the results with respect to program structure, the staff of one of the day care centers instituted some changes, thus allowing us to do a follow-up case study of a burnout intervention (Pines & Maslach, 1980).

The next set of survey studies (done in collaboration with Susan Jackson) were more systematic in their assessment of burnout. This was because they were part of our psychometric research program to develop a standardized measurement tool (a program described in the next section). In addition to generating needed psychometric data, each study was designed to test some specific ideas about burnout. For example, our study of police officers and their spouses (Jackson & Maslach, 1982) obtained independent spouse ratings, which provided evidence of convergent validity for our measure; however, it also tested some hypotheses about the relationship between burnout and home life. Other studies combined useful psychometric data with investigations of how burnout is related to critical job factors, demographic variables, and coping strategies (Maslach & Jackson, 1982, 1984b, 1985).

Psychometric Research

At this stage in our research, the key research issues were to develop a more precise definition of burnout and to develop a standardized measure of it. Our working definition of burnout consisted of two components: emotional exhaustion and depersonalization. Our next goal was to develop a standardized measure of this phenomenon, and so Susan Jackson and I spent the next few years conducting an extensive program of psychometric research (Maslach & Jackson, 1981a). We collected systematic data from hundreds of people in a wide range of health, social service, and teaching occupations.

Our findings confirmed the components of emotional exhaustion and depersonalization, but also revealed a third, separate component of feelings of reduced personal accomplishment. This empirically derived component was not inconsistent with the results of our earlier studies, but we had expected that such feelings would be one aspect of the other components and thus highly correlated with them. However, as a separate dimension, a feeling of reduced personal accomplishment is related conceptually to such phenomena as self-inefficacy (Bandura, 1977, 1982) and learned helplessness (Abramson, Seligman, & Teasdale, 1978).

On the basis of this psychometric research, we developed a measure called the Maslach Burnout Inventory (MBI). This measure was designed to assess the three components of the burnout syndrome: emotional exhaustion, depersonalization, and reduced personal accomplishment. There are 22 items, which are divided into three subscales. The items are written in the form of statements about personal feelings or attitudes, and are answered in terms of the frequency (on a 7-point scale) with which the respondent experiences them. In the original version of the MBI (Maslach & Jackson, 1981b), there was also a response scale for intensity of feeling; however, because of the redundancy between the frequency and intensity ratings, the intensity scale was deleted from the second edition of the MBI (Maslach & Jackson, 1986). The MBI has been used widely in research on burnout, and both the measure and the underlying multidimensional model have received strong empirical support (see Chapter 11).

MULTIDIMENSIONAL VS. UNIDIMENSIONAL MODELS OF BURNOUT

The popular, everyday view of burnout is that it is a single, unitary concept—and people either have it or they don't. This predilection to describe burnout in simple either-or terms is also reflected in the commonly expressed wish to be able to assess burnout with a measure that is short and produces a single score (with a clearly defined cutoff point). Certainly, it is easier and less complex to consider hypotheses about one dimension than several dimensions. It is not surprising, then, that the assumption of a single, unidimensional phenomenon underlies some conceptions of burnout (e.g., Freudenberger & Richelson, 1980; Pines & Aronson, 1988) and is reflected in the corresponding measures that have been proposed to assess it.

However, the empirical evidence provides more support for a multifaceted conception of burnout than it does for a single, unitary one. Our multidimensional model is not at odds with the simpler approach; rather, it both incorporates the single dimension (exhaustion) and extends it by adding two other dimensions: response to others (depersonalization) and response to self (reduced personal accomplishment). Interestingly, these three components have actually appeared within most of the various definitions of burnout, even if they have not been considered within a multidimensional framework. For example, exhaustion has also been described as wearing out, loss of energy, depletion, debilitation, and fatigue; depersonalization has been described as negative or inappropriate attitudes toward clients, loss of idealism, and irritability; and reduced personal accomplishment has been described as reduced productivity or capability, low morale, withdrawal, and an inability to cope. (For a more extensive analysis of these definitional issues, see Maslach, 1982c).

A clear trend in burnout research is that the variables studied so far are more strongly correlated with emotional exhaustion than with the other two components. This fact has led some people to argue that burnout is only the single dimension of emotional exhaustion, and that the other components should be dropped from the definition (e.g., Shirom, 1989). However, I disagree with this attempt to reduce and simplify the burnout concept. I suspect that one reason for the predominance of emotional exhaustion is that much of the theorizing about burnout focuses on that component rather than on the others. Thus, more of the research is explicitly designed to test variables related to emotional exhaustion and consequently it is not surprising to find more of the significant relationships on that dimension. The relationships of these variables to depersonalization and reduced personal accomplishment are often studied as an afterthought (and because it is easy to do the same statistical analyses with the other two subscales) rather than because of a well-articulated conceptual model. Clearly, more theorizing is needed with respect to the causes and consequences of these two components.

Of the three burnout components, emotional exhaustion is the closest to an orthodox stress variable. The factors hypothesized to relate to emotional exhaustion are very similar to those in the general literature on stress, and so the similar findings are not unexpected. Although this similarity validates the location of the burnout phenomenon within the stress domain, it is also the cause for some skepticism (Jackson, Schwab, & Schuler, 1986). If emotional exhaustion is simply a synonym for stress, then nothing new has been learned from the burnout research—it has simply replicated what was previously known under the guise of a new label (a charge that has been made by some critics). Thus, to limit the concept of burnout to just the component of emotional exhaustion is to define it simply as experienced stress and nothing more.

My own feeling is that there is indeed something more to burnout. To a large extent, this feeling is based on the information that emerged from the initial, exploratory research. However, it also reflects the general approach that I, as a social psychologist, have always taken to this phenomenon. Social psychology attempts to understand the behavior of the individual in a social context, and focuses on both social and personal factors. Thus, my social psychological analysis of burnout is that it is an individual stress experience embedded in a context of complex social relationships, and that it involves the person's conception of both self and others. To simply look at the stress component of this experience is not enough because it ignores the two latter components of self-evaluation and relation to others. In our three-component model of burnout, reduced personal accomplishment reflects a dimension of self-evaluation, and depersonalization tries to capture a dimension of interpersonal relations. Both of these components add something over and above the notion of stress; depersonalization, in particular, is a rather novel construct in the traditional job stress literature.

In their cogent analysis, Jackson et al. (1986) argued that significant progress in understanding burnout rests on the development of new, rather than traditional, theoretical perspectives. What is unique about the burnout syndrome, and what distinguishes it from other types of job stress, is what needs to be emphasized in future theoretical formulations. Otherwise, there will be no justification for the claim that it is a separate, meaningful phenomenon. Thus, an improved theory of burnout will need to develop an original model of all three of its components and not just the one of emotional exhaustion. This model should generate better hypotheses about both the social and personal causes and consequences of each of these components.

Moreover, this multidimensional model will need to articulate the interrelationships among the three components rather than simply considering each of them in isolation. Do these different dimensions develop simultaneously but independently, thus creating different patterns of burnout? Or are the dimensions related sequentially, so that there is a developmental progression over time? As an example of the first possibility, the pattern of the three components may be a meaningful index, as suggested by the empirical work on eight patterns, or phases, of burnout (Golembiewski, Munzenrider, & Stevenson, 1986).

An alternative approach postulates that there is a sequential progression over time, in which the occurrence of one component precipitates the development of another. There is some new theorizing and some initial empirical tests in this regard. The phase model (Golembiewski et al., 1986) argues that depersonalization is the first phase of burnout, followed by reduced personal accomplishment, and finally by emotional exhaustion. An alternative model was suggested by Leiter and Maslach (1988), in which emotional exhaustion occurs first, leading to the development of depersonalization, which leads subsequently to reduced personal accomplishment. Leiter (see Chapter 14) has amended that model to propose that personal accomplishment develops separately from the other two components (which are still linked sequentially). In other words, some burnout components develop in parallel rather than in sequence because they are reactions to different factors in the work environment.

Regardless of the specific form, a multidimensional model has some important implications for interventions. First, it underscores the variety of psychological reactions to a job that different employees can experience. Such differential responses may not be simply a function of individual factors (such as personality) but may reflect the differential impact of situational factors on the three burnout dimensions. For example, certain job characteristics may influence the sources of emotional stress (and thus emotional exhaustion) or the resources available to handle the job successfully (and thus personal accomplishment). This multidimensional approach also implies that interventions to reduce burnout should be planned and designed in terms of the particular component of burnout that needs to be addressed. That is, it may be more effective to consider how to reduce the likelihood of emotional exhaustion, or to prevent the tendency to depersonalize, or to enhance one's sense of accomplishment, rather than to use a more general stress reduction approach.

CONCEPTUAL IMPLICATIONS OF RESEARCH DIRECTIONS

As researchers, we always make choices about the way in which we study a particular issue—choices in terms of the questions asked, the theoretical frameworks applied, and the methodologies used. Such choices have important implications for what we do study, but also for what we do not. In my case, my social psychological background led me to frame the issue of burnout in terms of the social relationship between two people: one who gives, and the other who receives. What was most striking to me, particularly in the early stages of my research, was the nature of the interaction between these two people and the various factors that either enhanced or diminished that interaction. What was unique and intriguing about burnout (as opposed to other kinds of stress reactions) was the interpersonal basis for both sources of emotional strain and resources for coping. Clearly, my multidimensional approach grew out of that social psychological focus, particularly in terms of the two "additional" dimensions (depersonalization and reduced personal accomplishment) that place the exhaustion experience within a relational context.

One consequence of my social psychological framework has been that I have not done extensive research on the role of personality variables and individual differences in burnout. This is not to say that I think these variables are unimportant but rather that I think the social factors are the most critical ones to address. That may be a particular bias on my part, which may have had the (unintended) effect of directing other researchers away from the study of personality and burnout. The general finding that situational factors are more predictive of burnout than individual factors may reflect this bias rather than a true state of affairs with regard to burnout. In either case, I believe we need to rethink the entire issue of individual factors in burnout and try to develop a better conceptual model for hypothesizing which of these variables would be most relevant. The approach taken by Buunk and Schaufeli (see Chapter 4) seems particularly promising because their integration of the multidimensional model within a social comparison framework enables them to specify individual differences that are truly meaningful in terms of each of the burnout dimensions.

Another shortcoming of my social psychological approach has been that I have not conducted clinical research either on burnout symptomatology or on diagnostic criteria. As the coauthor of the MBI (which was designed specifically for research purposes), I am always being asked if it can be used for individual diagnosis as well. At the moment, the answer is no—but that could change if the appropriate clinical research were done. Certainly, the MBI has the potential to be used for diagnostic purposes, but without the necessary research there is no solid basis on which to identify meaningful cutoff scores or dysfunctional patterns of response. Thus, the development of such a diagnostic tool is an important research goal for the near future.

The development of the MBI itself has had a major impact on the shape and direction of burnout research. The availability of an easily administered, self-report measure was an attractive incentive to many researchers, who then incorporated the measure into their studies. In doing so, they accepted, at least implicitly, the multidimensional model on which the MBI was based. Thus, the availability of a research measure shaped the conceptual framework of many researchers, with the result that their studies were more method-driven than theory-driven (see also Chapter 1). The availability of the MBI has also contributed, I believe, to the current imbalance in research methodologies, i.e., to the overemphasis on one-shot, cross-sectional, self-report surveys. Such studies are certainly important, and can give useful information about large numbers of people, but they also have their limitations. Other types of research could fill in the gaps and provide new data that would supplement and extend what we can learn from surveys. Obviously, there are many reasons for the overreliance on self-report surveys; however, I think that the existence of the MBI may have been one of them. One of my concerns is that researchers begin to use it uncritically, or begin to think about burnout simply in terms of "correlated measures." For example, a question about the relationship between burnout and depression would be tested by administering the MBI and some depression measure, and then simply looking at the correlation between the measures. However, even if these measures are significantly correlated, these scores do not give us an understanding of why and how. To the extent that the use of self-report surveys removes the researcher from direct contact with the people being studied, it will become more difficult to gain a true understanding of the processes involved.

Oddly enough, although the earliest exploratory research was based on interviews and case studies (as in the work of Cherniss, Freudenberger, and myself), this approach has not been used extensively by subsequent researchers. In some ways, it is a more clinical approach, and not one used often by industrial/organizational (I/O) researchers; thus, its relative lack of use may reflect the greater I/O dominance that is now current in the burnout field. However, such an approach will certainly be a necessity in future research if we are going to gain a better understanding of the process of burnout and of its development over time. This approach will also be essential for expanding our thinking about the role of various social relationships in burnout.

SOCIAL RELATIONSHIPS REVISITED: A RETURN TO ROOTS

In general, I feel that the focus on the social interaction between the provider and the recipient has been lost in recent years as there has been an increasing emphasis on job factors and the use of I/O theoretical frameworks. The shift in label from "burnout" to "job burnout" is not an insignificant one. I believe that the addition of I/O theories and variables, as well as methodology, has been enormously useful and enriching for the study of burnout. However, it may have also shifted the focus away from the interpersonal, relational roots of burnout to the view that burnout is just another job phenomenon.

This job framework has led some researchers to extend the concept of burnout to many other occupations in which there is not an equivalent to a caregiving relationship or an ongoing interaction between people. Having done so, the researchers then say that "depersonalization" does not make sense in this job context, and therefore it should either be deleted from the definition of burnout or be transformed into another concept (see Chapter 1). The question here is whether "burnout" is actually the same phenomenon when it is transferred to these other occupations. Does it really make sense in those instances to continue to use the concept of burnout, or would it be better to conceive of those particular job issues in terms of some other, more appropriate construct? I must admit that I have had some doubts about the wisdom of viewing burnout as a general job phenomenon rather than a more specific one.

In a similar way, the view of burnout as a job phenomenon has led some researchers to question whether it is appropriate to use the concept in nonjob settings, as in the case of parent burnout or marriage burnout. From an I/O perspective, this may not make sense—but from my relationship perspective, it does. Indeed, I find it easier to translate the burnout concept into relationships other than the professional provider–recipient one than to translate it into occupations other than people-oriented ones. Whether this represents a limited vision on my part or an appropriate limitation on the concept has yet to be determined.

CONCLUSION

In my future work, I hope to make some new contributions to our understanding of burnout, but I want to do so again within the context of social relationships. I believe that there are many interesting questions that need to be addressed in relational terms, including issues of social power (and its imbalance in most helping relationships), interpersonal communication, attributions, and self-presentation, among others. This focus on relationships will of necessity involve a greater concern with process—which most theorists agree is essential for the further development of the burnout concept. Thus, my future direction represents not only a change from my most recent work but a return to my research roots, for it was my study of helping relationships that first led me to the discovery of burnout. It is my hope that these roots, when grafted with new material, will produce a new flowering of knowledge and understanding about the burnout phenomenon.

3

BURNOUT: AN EXISTENTIAL PERSPECTIVE

Ayala M. Pines
University of California, Berkeley

The introduction of burnout to the scientific community in the mid-1970s has been followed by several controversies among its scholars. One of the first major controversies at that time (and I will return to it later) centered on the question of definition. One of the major controversies today centers on the underlying dynamic of burnout.

In this chapter I will argue that the root cause of burnout lies in our need to believe that our lives are meaningful, that the things we do—and consequently we ourselves—are useful and important (Pines & Aronson, 1988). Frankl (1963) writes that "the striving to find meaning in one's life is the primary motivational force in man" (p. 154). When people try to find meaning in their life through work and feel that they have failed, the result is burnout.

What makes the search for meaning such a powerful force is the basic tragedy all human beings have to face, i.e., the finality and inevitability of death. According to Becker (1973), "The idea of death, the fear of it, haunts the human

The chapter benefited greatly from the comments of Christina Maslach and Wilmar Schaufeli.

animal like nothing else" (p. ix). Our need to believe that the things we do are meaningful is our way of dealing with the angst caused by facing up to our mortality. To avoid and deny death we need to feel heroic, to know that our lives are meaningful, that we matter in the larger, "cosmic" scheme of things.

How people choose to become heroes depends to a large extent on their culture's prescribed hero system. Whatever the hero system, according to Becker, people serve it "in order to earn a feeling of primary value, of cosmic specialness, of ultimate usefulness to creation, of unshakable meaning" (p. 5). In previous eras, religion was the most commonly chosen hero system, and for many it filled the purpose of transcending death admirably. Religious people knew the reason they were put on earth, i.e., to do God's will. For some people today a religious belief still provides this function. Cherniss and Krantz (1983) describe the nuns in a Catholic religious order who work in a residential setting for mentally retarded people. The nuns have no life outside of the religious order and their work with the retarded. They are in contact with their "clients" 7 days a week, yet despite the stress of their work maintain a high level of commitment and care for years. Because their work serves as a mean for achieving personal salvation, these nuns believe that they are getting back as much as they are giving.

Unfortunately, for the majority of people in our modern era religion is no longer adequate. For people who have rejected the religious answer to the existential quest, one of the frequently chosen alternatives is work. Thus the stakes have become very high. People who choose this alternative are trying to derive from their work nothing short of a sense of meaning for their entire life. When they think they have failed, they burn out.

Since religion provides a better answer to the existential dilemma than work (because God is infallible), people who attempt to find existential significance through their work are more likely to burn out than people who derive their existential significance from a religious belief. Following this line of reasoning, one possible interpretation of the flourishing of burnout these days is the secularization of society. A similar point of view has been forwarded by Lasch (1979) who talks about the failings of "the culture of narcissism" that has replaced the religious authorities of the past. More recently, Mander (1991) described what happens in "the absence of the sacred."

Failure in the existential quest for meaning is the root cause of burnout, as I will try to demonstrate throughout the rest of this chapter. This is why burnout tends to afflict people with high goals and expectations when entering such professions as pediatrics (Pines, 1981), nursing (Kanner, Kafry, & Pines, 1978; Pines & Kanner, 1982), organizational consultation (Pines, 1992) management (Etzion, Kafry, & Pines, 1982), kindergarten teaching (Maslach & Pines, 1977), social work (Pines & Kafry, 1978), and mental health work (Pines & Maslach, 1978). All this work suggests that when highly motivated professionals who identify with their work and hope to derive from it a sense of existential significance fail to accomplish their work goals and feel unable to make a significant contribution, they become susceptible to burnout.

When highly motivated nurses, who entered nursing "to help people . . . do good for humanity . . . make life and death more comfortable for fellow human beings" (Shubin, 1978), feel unable to relieve their patients' pain and suffering, they are unable to derive a sense of existential significance from their work and thus become susceptible to the danger of burnout. In a study of nurses (N = 32), the work characteristics that had the highest correlations with burnout were work pressure (the feeling that there is not enough time or sufficient force to do the work right) (r = .53; p < .001), a feeling of lack of success at work (r = .49; p < .001), and the sense of responsibility for patients' pain (r = .48; p < .001) (Pines & Kanner, 1982).

When "passionate, idealistic, and dedicated teachers" (Bloch, 1977) feel unable to educate and inspire their students because of apathy, discipline problems, overcrowded classrooms, shortage of available support staff, excessive paperwork, and excessive testing, they are likely to burn out (Farber, 1982). Indeed, when teachers' needs for self-actualization and self-esteem are unfulfilled, there is a high probability of burnout (Anderson & Iwanicki, 1981).

When psychotherapists whose motivations for entering the therapeutic profession were based on a wish "to help people and to gain gratification, joy and pleasure in the knowledge that [they] have been able to be of assistance to someone in need" feel that they have failed, that "no matter what [they] do, it won't be enough," they burn out (Freudenberger, 1983).

THE EXISTENTIAL PERSPECTIVE AND DEFINITIONS OF BURNOUT

As noted earlier, one of the early controversies among scholars studying burnout centered on the question of definition (e.g. Paine, 1982b; Maslach, 1982c). And indeed, the definitions were many and varied. Here are three of the most frequently quoted definitions. According to Freudenberger and Richelson (1980), burnout is "a state of fatigue or frustration brought about by devotion to a cause, way of life, or relationship that failed to produce the expected reward" (p. 13). According to Maslach (1982a), "burnout is a syndrome of emotional exhaustion, depersonalization, and reduced personal accomplishment that can occur among individuals who do 'people work' of some kind" (p. 3). According to Pines and Aronson (1988), burnout is "a state of physical, emotional and mental exhaustion caused by long term involvement in situations that are emotionally demanding" (p. 9).

While these definitions very in several aspects, all three of them (as well as virtually all the other definitions in the scientific literature) share a view of burnout as a state of fatigue and emotional exhaustion that is the end result of a gradual process of disillusionment. According to Freudenberger and Richelson, this process is caused by failure to produce an expected and desired goal. According to Maslach, the process is caused by doing "people work." According to Pines and Aronson, the process is caused by long term involvement in emotionally demanding situations.

Viewing burnout as the end result of a process implies an initial state of high motivation and high involvement. This is true for the three definitions of burnout presented above as well as for most other definitions found in the literature. For Freudenberger and Richelson, the initial state is characterized by devotion to a cause. For Maslach, the initial state is characterized by personalization and high personal accomplishment among individuals who do people work. For Pines and Aronson the initial state is characterized by high emotional involvement. In other words, in all three cases burnout is described as the result of a process of disillusionment that is typically found among highly motivated individuals. This process of disillusionment highlights once again the root cause of burnout—a sense of failure in the existential quest for meaning. If you don't feel a devotion to your cause, if you do people work but don't care about the people you work with, if you are not emotionally involved in your work—you are not likely to burn out. But if you are devoted to your work and are emotionally involved, if you expect to derive from your work a sense of existential significance—and you feel that you have failed— you are a likely candidate for burnout.

The existential perspective can be applied to virtually all of the descriptions of the process of burnout that appeared in the scientific literature. Let us examine, for example, three descriptions of the process of burnout that were provided by leading scholars in burnout research.

According to Cherniss (1980a, Chapter 8 of this book), a major source of burnout is professionals' inability to develop a sense of competence (1980a) and self-efficacy (Chapter 8). I would argue that feelings of efficacy and competence are so important because they give professionals a sense of existential significance. If my work makes a difference, I make a difference.

Excessive workloads, lack of administrative support, and bureaucratic constraints are stressful not only because systemic factors prevent professionals from using their skills in a way that would achieve their intended outcomes, as Cherniss (Chapter 8) argues. There is another, deeper reason. These stressors give workers a feeling that what they do is insignificant. Similarly, the experience of thwarted competence is stressful not only because achieving a sense of competence and success is important to workers, as Cherniss argues, but because thwarted competence deprives workers from experiencing a sense of significance in their work.

Clients who are resistant or do not improve cause stress to human service professionals not only because they prevent professionals from feeling competent and successful, as Cherniss argues, but because they prevent the professionals from achieving a sense of significance in their work. For the same reason, clients who improve and credit the professional with their improvement are considered the best and most rewarding clients. (See p. 50 for the exercise "Your Best and Your Worst Clients," which demonstrates this last observation.)

According to Maslach (1982a, p. 3), "A pattern of emotional overload and subsequent emotional exhaustion is at the heart of the burnout syndrome. A person gets overly involved emotionally, overextends him- or herself, and then feels overwhelmed by the emotional demands imposed by other people." I would argue that the person Maslach describes gets overly involved emotionally because that person doesn't look at the work as merely a job but wants to have impact on the people he or she works with, wants the work to make a difference in their lives, wants it to be significant. "Overly involved" means that the person identifies emotionally with the work, so that success at the work implies a personal success, while failure at the work implies a personal failure.

The emotional demands imposed by other people are not a problem as long as they can be satisfied. Actually, they are the challenge that makes the work seem important and significant. In my own work with human service professionals I have often heard: "It's not the actual work with people that burns me out. People are the primary reason I chose this career. It's my inability to help them that causes my burnout." The emotional demands imposed by people become a cause of burnout only when they are overwhelming, when it is impossible to respond to them adequately. Then they contribute to a subjective experience of failure. For the person who looks to find existential significance in work, such a failure is devastating and, as Maslach points out, a primary cause of burnout.

According to Freudenberger and Richelson (1980), burnout is the high cost of high achievement. When describing people who are prone to burning out, they say: "Initially, they enter the job market full of good intentions: idealistic, hopeful and somewhat naive. They give it their all and more, in order to attain the hoped-for good sense of self" (p. 179). I would argue that this good sense of self to which Freudenberger and Richelson refer is the sense that one's work efforts have an impact that matters, and that consequently one also matters. The good intentions, the ideals, and the hopes of these "somewhat naive" burnout-prone, high-achieving individuals are to have work achievements and accomplishments that make a difference in the world. Their success at work is expected to provide a sense of existential significance. That is why they "give it their all and more" (more than is necessary according to the formal job requirements) and that is why they burn out when they think they have failed.

Other descriptions of the burnout process that have appeared in the scientific literature can be analyzed similarly. In all cases the underlying dynamic appears to be the same, and at its core lies a sense of failure to find existential significance at work. As we will see later (after differentiating between burn-

out and other concepts with which it is frequently confused), this underlying
dynamic provides the foundation for the existential model of burnout.

BURNOUT AND OTHER CONCEPTS

Viewing burnout as the result of a failure in the existential quest for meaning
helps distinguish it from related concepts.

Burnout and Stress

Stress is defined by Selye (1956, 1982) as "the nonspecific (that is, common)
result of any demand upon the body, be the effect mental or somatic." The
formulation of this definition is based on objective indicators such as bodily and
chemical changes that appear after any demand. Lazarus distinguishes three
basic types of stress: systemic, psychological, and social (Monat & Lazarus,
1985). Systemic stress is concerned primarily with the disturbances of tissue
systems, psychological stress with cognitive factors leading to the evaluation of
threat, and social stress with the disruption of a social unit or system.

While everyone can experience stress, burnout can only be experienced by
people who entered their careers with high goals, expectations, and
motivation—people who expected to derive a sense of significance from their
work. A person who has no such initial motivation can experience job stress but
not burnout. Unlike stress, which can occur in endless types of situations (e.g.,
war, natural disaster, illness, unemployment), including endless types of work
situations, burnout occurs most often in work with people, and results from the
emotional demands that arise in the interaction with them. Stress does not
necessarily cause burnout. People are able to flourish in stressful and demanding jobs if they feel that their work is significant.

Burnout and Alienation

The prerequisite for burnout, the initial expectation to derive existential significance from work, also differentiates it from job alienation. Durkheim (cited in
Kanungo, 1979) describes alienation as a state of *anomie* that arises when people experience the lack or loss of acceptable social norms to guide their behavior and values. In later writings on alienation, sometimes scholars refer to the
reaction of a particular individual and sometimes scholars refer to a collective
(e.g., when a group of workers is said to be alienated); sometimes they refer to
certain objective conditions (e.g., the alienating properties of mechanization);
and sometimes to subjective psychological states experienced by an individual
or a group (Kanungo, 1979; Monat & Lazarus, 1985). While people who are
burned out in their job often feel alienated, they did not feel that way initially.
Alienation is a general experience that can occur in people who have never

expected anything from their work except a paycheck. (Assembly line workers are prone to experiencing alienation but not burnout.) Burnout most often happens to people who initially cared most about the people they work with and least about their paychecks.

Burnout and Depression

Burnout is different from clinical depression. According to the DSM-III, depression is a dysphoric mood or loss of interest in or pleasure derived from all or almost all activities and pastimes. The dysphoric mood is described as depressed, sad, blue, hopeless, low, down in the dumps, irritable, and so forth.

Unlike burnout, depression tends to be all-pervasive. At the early stages of job burnout, people are often still happy and productive in other spheres of life. Despite the tremendous advances in psychotherapy for depression (especially cognitive therapy), pharmacotherapy remains the standard treatment (Beck & Young, 1985). During treatment for depression, the individual's psyche and early childhood experiences are viewed as the source of symptoms and the focus of therapy. In burnout, on the other hand, except for the most extreme cases, the search for both causes and the cures focuses on stresses in the work environment and on people's need to derive a sense of existential significance from their work.

Burnout and Existential Neurosis (Existential Crisis)

Maddi (1967, 1970) describes existential neurosis as a "chronic inability to believe in the truth, importance, usefulness or interest value of any of the things one is engaged in or can imagine doing." It is the feeling that one has nothing to live for, nothing to hope for. Camus (1955) states, very much like Frankl (1963), "I have seen many people die because life for them was not worth living. . . . From this I conclude that the question of life's meaning is the most urgent of questions" (pp. 3-4). According to Yalom (1980), the question of life's meaning can take many forms: "What is the meaning of life? What is the meaning of *my* life? *Why* do we live? *Why* were we put here? What do we live *for*? What shall we live *by*? If we must die, if nothing endures, then what sense does anything make?" (p. 419).

For most burned-out professionals, work initially provided (or was expected to provide) an answer. They knew why they were put on earth: to do the work for which they had a calling. Burnout characterizes people who start out believing that the work they do is important, caring deeply about the people they chose to help, and hoping to have a significant impact on their lives and make the world a better place to live in. They burn out when they feel that they have failed.

Burnout and Fatigue

One major difference between burnout and physical fatigue is that one can recover quickly from the latter but not from the former. In addition, while people who are burned out also feel physically exhausted, they describe it as very different from the "normal" experience of physical fatigue. Physical effort and strenuous physical exercise cause fatigue, but such fatigue is usually experienced positively and is accompanied by feelings of accomplishment and success. Burnout, on the other hand, is a negative experience that is accompanied by a deep sense of failure.

Conclusion

Taken as a whole these concepts seem to suggest that the unique characteristic of burnout, the one that best differentiates it from the other concepts (including job stress, fatigue, alienation, depression, and existential crisis), is the fact that burnout is always the end result of a gradual process of disillusionment in the quest to derive a sense of existential significance from work.

While it may be true that in the discussion of depression, for example, there is an implicit assumption that the depressed individual may not have always been depressed, the state of depression and not the sinking down to it is the crucial element in both its conceptualization and its treatment. The same thing can be said about all the concepts I have chosen to address in this chapter as well as the other concepts that have been presented at one time or another as being "the same as burnout."

Another major difference between burnout and these related concepts is that burnout is a much more specific phenomenon. While stress and fatigue happen to everyone, and can occur as a result of an endless number of situations, burnout happens only in people who entered their professions with an expectation to derive from it a sense of existential significance. Similarly, while depression, alienation, and existential crisis are total and general experiences that have an impact on all aspects of a person's life, burnout is a specific experience that characterizes people who work over long periods of time in situations that are emotionally demanding—the kind of situations that occur frequently in work with people.

After demonstrating the applicability of the existential perspective to all the major definitions of burnout, and after demonstrating the value of the existential perspective in differentiating burnout from other related concepts, we can move on to a discussion of the existential model itself. As an introduction, I would like to present two related models of burnout.

AN EXISTENTIAL MODEL OF BURNOUT

In his chapter on the role of professional self-efficacy in the etiology of burnout (Chapter 8), Cherniss quotes Hall's concept of "psychological success" (Hall, 1976). According to Hall, work motivation and satisfaction are enhanced when a person makes a commitment to a goal, and through independent effort—which requires autonomy and support—attains that goal. The goal should be "personally meaningful." Such achievement leads to a sense of psychological success, which leads to increased self-esteem, which leads to increased job satisfaction, which in turn encourages the individual to become more involved in the job. While Hall does not discuss burnout directly, his model was influential in shaping Cherniss's idea about the crucial link between burnout and the inability to achieve a sense of competence in one's work.

Harrison (1983) offers a model of burnout that is similar in several important elements to the model proposed by Hall and Cherniss. It is a model of "social competence." It starts with the motivation to help that characterizes people who enter the social services. When the job provides "helping factors" (unambiguous expectations, resources, information, etc.) people are able to achieve high effectiveness in their work, which provides them with a sense of competence, which in turn enhances their original motivation to help. When the work experience is primarily of "barriers," the result is low effectiveness, which causes burnout, which reduces the motivation to help.

Hall's psychological success, Cherniss's feelings of competence and success, and Harrison's sense of competence and feeling of efficacy are all important parts—but only parts—of the existential sense of significance that professionals can derive from their work. The reason why success, competence, and efficacy are so very important is that they provide professionals with a sense that the things they do—and consequently they themselves—are useful and important, that their lives matter in the larger scheme of things. This view is represented in the existential model of burnout (see Figure 3-1).

Motivations

Like the other two models, the existential model of burnout is motivational. Its underlying assumption is that only highly motivated individuals can burn out. In other words, in order to burn out, one has to first be "on fire." A person with no such initial motivation can experience stress, alienation, depression, an existential crisis, or fatigue, but not burnout. The view of the critical role played by the initial spark in the burnout process is shared by other scholars. Harrison (1983), for example, states that "practically all beginners are in some sense 'fired up' with motivation" (p. 32). And Farber (1983b) concludes that "a high level of commitment to one's work is often regarded as a prerequisite to burnout" (p. 9).

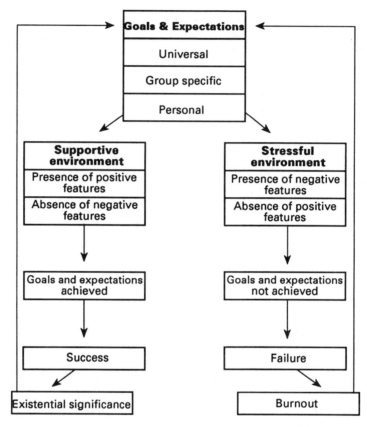

FIGURE 3-1 Existential model of burnout. Adapted from Pines and Aronson, 1988, p. 34.

According to the existential model of burnout, highly motivated individuals enter their chosen career with high goals (for themselves) and high expectations (of what the work will provide). Some of these work-related motivations are universal and shared by most people when they enter the work force. Some of the motivations are profession-specific and shared by the group of people who choose a particular profession. Some are personal and unique to the individual worker (Pines, 1982b).

The *universally shared work motivations* include such goals and expectations as: to have significant impact, to be successful, and to be appreciated. For example, one study involving 205 professionals from a wide range of occupations revealed that burnout was negatively correlated with a sense of success ($r = -.48$), with the ability to express oneself on the job, and with the level of appreciation received for the work done (both $r = -.31$) (Research Appendix, Pines & Aronson, 1981).

The *profession-specific work motivations* reflect a "profile" that character-izes people in a particular field of work. This profile is the result of two interrelated processes: selection and professional socialization. In terms of se-lection, the nature of the occupational tasks acts as a screening device, attract-ing people with particular motivations. Once these individuals enter a career, the process of professional socialization teaches them values and behaviors viewed as "appropriate" for their particular professional role.

While every profession has its own unique set of goals and expectations, there are certain work motivations that are shared by most people in the human services. As Harrison (1983) notes, "The overwhelming majority of people who enter the social services are highly motivated to do something good *for*, and sometimes *with* other people" (p. 31). And Freudenberger and Richelson (1980) state similarly: "The helper has come to his (or her) profession with visions of . . . ability to create a difference in people's lives" (p. 154). Clearly, for people who choose a career in the human services, the shared work motiva-tion is to help people and have a significant impact on their lives. In virtually every burnout workshop I have ever conducted with human service profession-als (Pines, 1992), when I asked participants about their goals and expectations when they entered their profession, the shared motivation was to work with people and to have a significant impact on them. Kadushin (1974) termed this motivation a "dedicatory ethic," which elevates service motives so that the work is no longer seen just as a job but rather as a calling.

The *personal work motivations* are based on an internalized "romantic image" of the work that is modeled after an important person, an admired character in a book or a movie, or a significant event. For one prison psycholo-gist the role model was a dedicated prison psychologist who had written about his work years earlier. For one nurse the role model was the loving nurses who took care of her when as a child she was hospitalized for injuries suffered in a car accident.

Combined, these universal, profession-specific, and personal motivations form an expectation that the work will have significant impact and will enable the individual to be successful (be a hero). Whether this expectation will come true depends to a large extent on the work environment.

Work Environments

When in a supportive work environment, highly motivated individuals can usu-ally achieve their goals and expectations. A supportive work environment is one that provides a maximum of positive features that enable professionals to reach their goals by providing the needed autonomy, resources, and support. It also minimizes negative features that can interfere with goal attainment. In six dif-ferent samples (total $N = 1,827$) burnout was found to be significantly and negatively correlated with autonomy (r ranging from $-.15$ to $-.32$), variety (r

ranging from $-.14$ to $-.35$), facility effectiveness (r ranging from $-.18$ to $-.30$), policy influence (r ranging from $-.15$ to $-.18$), work sharing (r ranging from $-.22$ to $-.28$), work relations (r ranging from $-.23$ to $-.32$), social support (r ranging from $-.17$ to $-.29$), and feedback (r ranging from $-.15$ to $-.36$). In an effective organization, professionals have the political power and autonomy to do the work as it should be done, feel challenged and supported, have good relation with other workers, and consequently feel successful. Indeed, the correlations between burnout and perceived success in these six samples ranged between $-.15$ and $-.48$.

Success provides a sense of existential significance that in turn reinforces these individuals' original motivation for the work. This positive loop can be sustained indefinitely, as long as the individual feels challenged, supported, and successful at work. When the same highly motivated individuals enter a stressful work environment, an environment in which obstacles and unmodifiable stresses are maximal while rewards, support, and challenge are minimal, they cannot get the opportunity, resources, or authority they need to achieve their goals. In two samples (total $N = 929$) burnout was found to be significantly and positively correlated with such work features as overextension ($r = .22$ and $.31$), overload ($r = .13$ and $.35$), decision load ($r = .19$ and $.30$), guilt about not providing adequate service ($r = .29$ and $.42$), environmental pressures ($r = .27$ and $.21$), bureaucratic pressures ($r = .20$ and $.24$), administrative hassles ($r = .20$ and $.26$), social overextension ($r = .16$ and $.38$), and conflicting demands ($r 1 = .27$ and $.31$). The bureaucratic and administrative pressures limit the professionals' freedom at work and force them to spend time and effort on things they consider secondary in importance. Overload, pressure to make important decisions without adequate time, conflicting demands, and overextension give them the feeling that no matter how hard they try, they can never have significant impact or be successful. The guilt over not providing the people they work for with adequate service is extremely stressful.

The result of a long-term confrontation with these negative work features is a subjective experience of failure. For people whose egos are tied to their performance, and who are trying to derive from their work a sense of existential significance, failure is a devastating experience and a powerful antecedent of burnout.

It is not always and not necessarily an objective failure that causes burnout, but rather the feeling that no matter how hard one works one can never have a significant impact. A social worker who was treating multiple problem families, for example, burned out when she realized that even if she worked 12 hours a day, 7 seven days a week, she could never have a real impact on those families' lives. In her organization she was considered a very successful professional, but in her own mind she was a failure because she felt that all her efforts were insignificant and meaningless. The frequently noted observation that professionals are rarely fired because of burnout but often choose to leave their career as a result of burnout supports this contention.

Once burnout starts, it reduces the individual's motivation for work. The result is a negative loop that with time and with growing levels of burnout turns some people into "dead wood," makes some people quit their job, makes other people go back to school so they can climb the administrative ladder and escape the emotionally demanding work, and causes others to leave their chosen careers altogether.

The crucial factor in determining whether highly motivated individuals will achieve a sense of success and significance or whether they will feel that they have failed and burn out is the perceived work environment. While it is true that sometimes people can feel successful even in a stressful environment, and can feel that they have failed in a supportive environment, these are the exceptions and not the rule. In most cases these exceptions can be traced back to individual goals and expectations that are unusually high or low. In the majority of cases, however, environmental conditions play the major role in determining whether highly motivated individuals will burn out or will feel that the work they do is meaningful and significant.

It is important to remember that the existential model of burnout is an abstraction. Real-life work environments are never all-supportive or all-stressful; rather they consist of a complex combination of supportive and stressful features. The likelihood of burnout occurring depends on the key factors in the balance between the supportive and stressful elements in the environment.

Even when a particular work environment is very stressful, if the work provides a sense of success and significance, people will not burn out. A relevant example mentioned earlier is the common report by human service professionals that despite the emotional stress involved in client contact, it is not the primary cause of their burnout, because contact with clients is also the most significant aspect of their work and the reason they chose that work to begin with.

A moment's reflection makes it clear that the environment—like all other elements in the model—is to a large extent subjective (or, to borrow Hall's terminology, "psychological"). Overload is a negative environmental feature, yet the same amount of work can be overload for one person but not another. Challenge is a positive feature, yet the same task can be challenging for one person but not another. Consequently in the same environment one person may feel successful and another may not.

The fact that people respond to their subjective environment does not mean that their environment is totally unrelated to the objective reality. When the majority of professionals in a particular organization feel overextended and stressed by such things as overload, decision load, environmental pressures, bureaucratic pressures, administrative hassles, and conflicting demands, we can assume that there is a shared reality to which they are responding. The same professionals may feel differently in another organization where they are given more autonomy, variety, political influence, support, feedback, and appreciation.

Nevertheless, even if the difference between these two imaginary organizations in overall burnout rate is statistically significant, there will no doubt be sizable differences in mean burnout between different people within each organization. These individual differences reflect the subjective component of the environment that is the result of an interaction between the individual and the objective world. That interaction is at the heart of the burnout model.

Individual differences in burnout can usually be traced back to differences in the individual goals and expectations (the "romantic image") with which people have entered their chosen careers. The higher and the more unrealistic these goals and expectations are, the more likely they are to lead the individual to an experience of failure and eventually to burnout. While such individual work motivations are important, they are only one part of the motivation with which people enter work. As noted earlier, the other parts are the universal and the profession-specific motivations.

Finally, it is important to note that the existential model of burnout is not static. People can move from one part of the model to another at any time. Even when one works in a generally supportive environment, one may at one time or another fail to achieve an important goal, experience failure and with it the first signs of burnout. Similarly, even when one works in a stressful and nonsupportive work environment, one can at some point achieve an important goal, experience success, and derive from that success a sense of existential significance. Furthermore, even when a cycle of burnout has started, it can be interrupted at various points (e.g., by a reaffirmation of the personal or group-specific goals).

The remainder of the chapter will focus on coping with burnout. Two experiential exercises inspired by the existential perspective will be presented as a demonstration of the value of the model to the understanding and treatment of burnout.

COPING WITH BURNOUT

In this section two exercises are disscused that are inspired by the existential perspective.

Exercise 1: "Expectations and Stresses"

This exercise provides an experiential demonstration of the burnout model. As a result of this exercise, workshop participants discover the crucial link between their original goals and expectations and those elements in their work environment that now contribute to their burnout (Pines, 1992).

I begin the exercise by asking participants to write down the goals and expectations they had when starting their career. This helps to focus their attention on their original motivation for choosing their career. This is the starting point for the model of burnout and the possibility of their own burnout. Next, participants are divided into groups of four. The people in each foursome are asked to share with each other their goals and expectations, and to discover which of these are shared by all four of them. This enables them to differentiate between their universal and group-shared goals on the one hand, and their personal goals, the ones related to their romantic image of their chosen career, on the other hand. The information about the shared goals and expectations is presented by a chosen spokesperson and written on a board. When the entire workshop group is from one field of work, the shared goals show both the universal and the group-specific motivations.

These shared goals and expectations reflect the sense of significance that participants originally expected the work to provide. The universal motivations that can be found in practically every group include such things as: to do interesting and significant work, to be appreciated, and to be successful. For people in the human services there is always the item: to work with people. And for specific professions there are specific motivations. For nurses, this always includes the goal of easing pain; for teachers, to educate and enlighten; for social workers, to help those in need; for managers, to be able to run a department or organization effectively and the way it should be run. The generation of the list of goals and expectations through this process demonstrates to the participants that the burnout model is relevant to their own individual and profession specific case.

After discovering how they hoped to derive a sense of existential significance from their work, the workshop participants move to the next part of the burnout model and analyze those features in the environment that make it burnout inducing. This time they are asked to list what are at present the features in the environment that most contribute to their burnout. Then they return to their foursomes, share with each other the causes of their burnout, and discover the burnout-causing stressors they all have in common. The shared stressors are once again presented to the whole group and written on the board.

Typically, both the list of goals and the list of stressors have about 10 items each, whether they have been generated by a group of 8 or a group of 500. Once these two lists are written on the board, it becomes evident that most of the items are shared by all of the participants. Not surprisingly, different groups from the same profession generate very similar lists of both goals and stressors, and are fascinated to hear the other groups' lists. The recognition of this similarity and the universality of the experience is extremely important in coping with burnout because it helps break the fallacy of uniqueness (the false assumption that "I am the only one experiencing burnout").

In the third stage of the exercise, participants are asked to look at the two lists on the blackboard and note if there is any relationship between them. What becomes abundantly clear is that the burnout-causing stressors can be stated, in almost every case, as frustrated goals and expectations. Nurses find most stressful those aspects of hospital work that make it difficult to ease the pain of sick people. Teachers find most stressful the discipline and budget problems that make it difficult for them to educate and inspire students. Social workers find most stressful the bureaucratic aspects of the work that make it difficult to help people in need. Managers find most difficult the administrative restrictions and bureaucratic obstacles that make it difficult to run their organization effectively.

Participants are also encouraged to examine the stressors on their personal list that they did not share with the others in their foursome or the rest of the group. Often they discover that these stressors are related to a disappointment in an aspect of work that is related to a romantic image held when first embarking on their career.

The comparison between goals and stressors shows in a personally relevant way that the most burnout-causing aspects of work are those that prevent people from achieving their goals and expectations. This connection is represented in Figure 3-1 by the bar "Goals and expectations not achieved." The exercise offers participants the opportunity to discover for themselves the connection suggested in the model between their inability to achieve their goals, their painful feelings of failure, and their symptoms of burnout. Furthermore, they realize that their perceived failure is so painful because what they aspired for was of major importance: to derive a sense of existential significance from their work.

Once the connection between their goals and their burnout-causing stressors has been established, there is usually a visible sign of relief in participants as they stop blaming themselves for their burnout. The energy freed by this relief is then used to brainstorm creative ideas regarding how to change the work environment (or in some cases the participants' own unrealistic goals and expectations) in order to make the work more rewarding.

Illustration: Expectations and Stressors in Crisis Intervention

To illustrate the outcomes of this process, let us first examine the list of goals and expectations generated during a burnout workshop with crisis intervention counselors:

- Help people in crisis
- Work with people
- Do something significant
- Have interesting, emotionally satisfying work
- Be a part of support group of colleagues
- Actualize self
- Receive from giving
- Understand self better

- Continue to grow professionally
- Get to know different types of people
- Learn crisis intervention tools

The stressful, burnout-causing aspects of crisis intervention work mentioned most often were the following:

- Chronic cases
- Suicide cases
- Feelings of responsibility after a difficult encounter
- Feeling stuck in a difficult case
- Lack of feedback
- Lack of collegial support
- Professional isolation
- Lack of opportunities for professional growth
- Bureaucratic interference

When asked whether there was any relationship between the list of their goals and expectations and the list of their burnout-causing stressors, participants noted the relationship right away. Every stressor could have been described as a frustrated goal or expectation:

- Chronic cases: frustrate the expectation for interesting and emotionally satisfying work
- Suicide cases: prove that the counselor is unable to help people in a serious crisis
- Feelings of responsibility after a difficult encounter: frustrate the expectation to receive from giving
- Feeling stuck with a difficult case: proves that the counselor is unable to do something significant
- Lack of feedback: disappoints the goal of learning crisis intervention
- Lack of support from colleagues: disappoints the expectation to be a part of support group of colleagues
- Professional isolation: disappoints the expectation to be working with people
- Lack of opportunities for professional growth: disappoints the expectation for self-actualization and continuous professional growth
- Bureaucratic interference: makes it difficult to do something significant

Counselors tend to be humanitarians whose primary goal is to help people in trouble. They tend to value themselves most as being empathic, understanding, and helpful. They expect helping people in crisis to give their life a sense of meaning. A woman in the workshop expressed this best when she said: "When

I succeed in helping a person in a crisis, I feel that there is a justification for my existence."

Each one of the burnout-causing stressors mentioned by the counselors reflects a failure to achieve existential significance in their work. "Chronic cases" are people who use the crisis intervention center repeatedly, who are "stuck" in terrible life circumstances yet don't follow the counselors' advice. These chronic cases give the counselors a feeling that nothing they do helps, that their work has no impact. It is easy to see why for people whose major goal in work is to help people such chronic cases are a cause of frustration and a feeling of being exploited. Clients who commit suicide seem very different from chronic cases, but actually they are stressful for a similar reason: they represents the ultimate failure for the counselor.

Feelings of responsibility are stressful after a difficult encounter, but they are not stressful after a successful encounter. The reason? After a difficult encounter the professional feels responsible for the negative outcome of the session but powerless to bring about change. A similar stress is involved in feeling stuck with a difficult case: having no impact. "I didn't succeed in helping. Maybe I'm incapable of helping." Lack of feedback is stressful because the counselor doesn't know whether or not the intervention has been effective. Lack of collegial support and lack of opportunities for professional growth add to the feeling of professional isolation and being "stuck." Bureaucratic demands make counselors feel that they are dealing with trivia rather than with the "real" thing: helping people in trouble.

Exercise 2: "Your Best and Your Worst Clients"

Another demonstration of the existential perspective on burnout in a workshop context involves professionals' perceptions of their best and worst clients. In the first stage of this exercise, I ask workshop participants to think about their two most favorite clients (or employees or service recipients), the two they most enjoy working with. They then make a list of the traits these two people have in common. These traits usually include openness, intelligence, positive energy, appreciation, and the ability to grow and improve. I ask participants to call out some of the items on their list. Those items are written on the board under the heading "best client." Once about 10 traits are written on the board, participants indicate by a show of hands who agrees that each trait also describes their own best clients. They discover that their best clients are not unique. Rather, there are common traits that describe the universal best client. The same process is used to identify the traits of the universal worst client.

Once the trait lists of the best and worst clients have been identified and voted on, participants are asked to compare the two lists and note if there are any dimensions that the best and worst clients share. They inevitably discover that there are indeed such dimensions. If their best client is intelligent, their worst client is dumb; if their best client exudes positive energy, their worst client is negative and depleted of energy; if the best client is appreciative, the worst client is unappreciative; if the best client is able to grow and improve, the worst client is unable to change; if the best client is open, the worst client is closed-minded.

In the next and last stage of this exercise, participants are asked to contemplate whether there is any relationship between the shared dimensions of the best and worst clients and their own goals and expectations from their work. They of course realize that there is a connection. The best clients are those who enable the professionals to achieve their goals and feel that they are successful, that their work is important and has a significant impact. The worst clients are those who prevent them from achieving their goals and contribute to their feelings of failure. In other words, the most important dimension is the dimension of significance. We perceive and evaluate the people we work with according to the degree to which they fulfill our need to derive existential significance from our work.

In both of these exercises, realizing the role played by lack of significance in the etiology of burnout has an important implication for coping. Once participants understand this crucial link they can focus their efforts on ways to derive a sense of significance from their work and thus avoid burnout. Very often the result is a renewed sense of significance and with it a reduced level of burnout.

CONCLUSION

Burnout is a negative state of physical, emotional, and mental exhaustion that is the end result of a gradual process of disillusionment. It is typically found among highly motivated individuals who work over long periods of time in situations that are emotionally demanding. In this chapter I have tried to argue that the most emotionally demanding aspect of a work situation is its lack of existential significance. People need meaning in their lives, and the failure to find such meaning will cause burnout. It is not objective failure per se that causes burnout but rather the feeling that one's efforts are insignificant and meaningless. Similarly, it is not objective success per se that prevents burnout but rather the subjective experience of doing something meaningful.

The existential perspective on burnout has both practical and theoretical implications. On the practical level, by focusing on the importance of deriving a sense of significance from work it can guide attempts to cope with and prevent burnout. On the theoretical level, the burnout model makes a number of predictions that can be tested in future research. Both of these directions can help further the theoretical development of burnout.

4

BURNOUT: A PERSPECTIVE
FROM SOCIAL COMPARISON THEORY

Bram P. Buunk
University of Groningen

Wilmar B. Schaufeli
University of Nijmegen

Despite the fact that occupational burnout in the human service professions has been the focus of numerous research efforts, most research in this area has been atheoretical and has focused little attention on the social psychological processes that might be relevant. Moreover, although in-depth social psychological analyses of the burnout phenomenon have been presented (e.g., Harrison, 1983; Maslach & Jackson, 1982), these approaches have lacked firm empirical evidence. This chapter tries to bridge the gap between social psychological theory and burnout research. This will be done in part by linking burnout to recent developments in social exchange theory (e.g., Buunk & VanYperen, 1991; Walster, Berscheid, & Walster, 1978). However, the main focus in this chapter will be upon applying recent theoretical work on social comparison processes under stress to occupational burnout (e.g. Taylor, Buunk, & Aspinwall, 1990; Taylor & Lobel, 1989).

Our central thesis is that burnout develops primarily in a social context, and that to understand the development and persistence of burnout attention has to be paid to the way individuals perceive, interpret, and construct the behaviors of others at work. Two major assumptions behind our perspective are the following. First, individuals in the human service professions are involved in relationships with clients and patients, and in these relationships social exchange processes and expectations of equity and reciprocity play an important role. As Maslach (1982b) noted, a characteristic of burnout is that the stress arises from the social interaction between helper and recipient. Second, individuals will be inclined to deal with problems at work by engaging in social comparison with their colleagues and superiors, and by relating their own experiences to those of others—particularly colleagues in similar positions. Such comparisons may have consequences for the development and persistence of burnout symptoms.

We will first discuss recent developments in social comparison theory that are important for understanding stress at work. Next we will describe a number of major stressors in the nursing profession and a number of personality variables that seem relevant to burnout. We will then present some findings from a study among nurses, and we will show that each of the burnout dimensions proposed by Maslach (1982b) has different relationships to various stressors and personality characteristics. Finally, we will discuss some of our findings on the role of social comparison processes as related to burnout.

SOCIAL COMPARISON THEORY, STRESS, AND AFFILIATION

Research on stress and social comparison was originated by the classic experiments of Schachter (1959) on the relationship between fear and affiliation. Although Festinger (1954) confined himself to the evaluation of abilities and opinions, Schachter expanded the domain of social comparison to include emotions as well. His research showed that women who were experiencing fear because they were anticipating some electric shocks wanted to be with someone else, but only someone who was in the same situation. According to Schachter, individuals under stress seek out others for reasons of self-evaluation in order to assess the appropriateness of their own reactions. Later research substantiated this idea by showing that the need for social comparison is enhanced when individuals feel uncertain about how to feel and react. This was, for example, demonstrated when uncertainty was manipulated by false feedback (Gerard, 1963) and when the source of one's arousal was unknown (Mills & Mintz, 1972).

Over the past decade, there has been a resurgence of interest in social comparison tendencies under stress, particularly stimulated by Wills's (1981) influential paper. Wills suggested that when individuals are confronted with a threat to self-esteem, they engage in downward comparisons with less competent others in an attempt to restore the way they feel about themselves. This motive is called self-enhancement. Downward comparisons may, according to

Wills, lead to derogation of others or to affiliation with less fortunate others. Indeed, a number of survey studies have shown that individuals faced with serious diseases and crises, such as arthritis patients (Blalock, McEvoy-DeVellis, & DeVellis, 1989), mothers of medically fragile infants and women with impaired fertility (Affleck & Tennen, 1991), and cancer patients (Wood, Taylor, & Lichtman, 1985), tend to compare themselves with others who are worse off, and to perceive themselves as better off than most others facing the same or a similar stressor. For instance, among mothers of high risk infants, most mothers mentioned some aspect in which they felt better off than other parents with such infants (Affleck, Tennen, Pfeiffer, Fiflied, & Rowe, 1987). Among arthritis patients, the perception that one had fewer problems with one's performance than other patients made patients more satisfied with their own performance (Blalock et al., 1989).

As noted by Taylor et al. (1990), the focus in research on social comparisons under threat has been quite different than was the case in the original work of Schachter (1959) and in subsequent experimental studies on fear and affiliation (Cottrell & Epley, 1977). Schachter emphasized the affiliative activity that occurred in response to threat, but most recent research has focused on cognitive social comparison activity. Such activity constitutes the bringing to mind of other people as a way of making downward comparisons, and is characterized by the self-serving perception and construction of others as being worse off. It is important to make a clear distinction between this last process and affiliation because there is increasing evidence that in stressful situations there is no preference whatsoever for downward affiliation. Thirty years ago, Rabbie (1963) showed that the high-fear person was avoided in all experimental conditions. In another study, cancer patients indicated a preference for affiliation with others who were similarly or better off, although more subjects preferred someone similar (Molleman, Pruyn, & Van Knippenberg, 1986). Individuals with problematic marriages preferred on the average contacts with those who had better marriages, but those with happy marriages indicated a preference for contact with others who were as happy as they were (Buunk, VanYperen, Taylor, & Collins, 1991). Other evidence indicates that interaction with depressed individuals is aversive and leads to the desire to avoid further interaction (Coyne, 1976a).

As Gibbons and Gerrard (1991) noted, the foregoing suggests that while a person may find solace in the realization that other people are struggling even more with the same problems, he or she does not necessarily want to be in the presence of those people. In a similar vein, Taylor and Lobel (1989) suggested that people under threat avoid contact with persons who are doing worse, or are worse off, and prefer actual contact with persons who are doing better. Taylor and Lobel argue that individuals under stress are faced with two major coping tasks: regulating their emotions and obtaining relevant problem-solving information (Lazarus & Folkman, 1984). The first need is best addressed through the use of downward, self-enhancing comparisons, but the latter requires affiliation with, or information about, people who are better off. Thus, in this model upward affiliation serves the motive of self-improvement. Such contact may provide a person with valuable information for potential long-term survival and successful coping; constitute a method for obtaining hope, motivation and inspiration; and enhance the person's self-efficacy (see Chapter 8). It must be noted that Taylor and Lobel assume that individuals under stress not only desire to affiliate with those who are better off but wish to obtain information about such others as well.

STRESSORS IN THE NURSING PROFESSION

Social comparison theory seems particularly relevant for understanding burnout among nurses because, as in many other human service professions, *uncertainty* seems a rather salient stressor within nursing, and uncertainty is supposedly a major factor instigating social comparisons (Buunk et al., 1991; Molleman et al., 1986). The concept of uncertainty as we employ it refers *not* to ambiguity about the environment but to lack of clarity about what to feel and think, or how to act. Although the role of the nurse may seem clear, there may be considerable uncertainty as to how to carry out this role (e.g., McGrath, Reid, & Boore, 1989). For example, nurses may wonder if they are too involved with patients or not involved enough, may feel uncertain about how to deal with various problems of patients (including appeals for help and expressions of anxiety), and may experience insecurities about whether they are doing things right. Cherniss (1980a, pp. 206–212) considers doubts about competence to be a major source of stress that can lead to burnout in human services professionals, particularly in the early stages of their careers. In the same vein, Gray-Toft and Anderson (1981) found that inadequate preparation and uncertainty concerning treatment were among the severest stressors in nursing.

A second stressor that seems prominent in the nursing profession is *imbalance* between investments and outcomes in relationships with patients. The notion behind this stressor is based on social exchange theory (cf. Walster et al., 1978). The assumption is that there exists a characteristic human tendency to expect some reward such as gratitude in return from others to whom we provide caring, empathy, and attention. But within the health professions, such expectations are often not fulfilled (cf. Maslach, 1982b; Maslach & Jackson, 1982). Patients may be worried and anxious, and interactions with such individuals may not be rewarding. An additional problem may be that patients often do not follow advice or guidelines, and may therefore improve only slowly or not at all. As a consequence of such processes, nurses may often feel that what they invest in their relationships with patients is not in proportion to what they get out of them. Although being bothered by the lack of equity in such relationships seems to contradict the dedicatory ethic (Kadushin, 1974) characteristic for caregivers, an imbalanced relationship may put considerable pressure on a nurse.

The last stressor is *lack of control*, a variable that plays a central role in most stress theories, is generally acknowledged to affect mental health and well-being of people in organizations (Ganster, 1989), and has been linked to negative impacts of social comparisons (e.g., Major, Testa, & Bylsma, 1991). For nurses, many aspects of the work environment are beyond their control, including the recovery of patients, decisions made by physicians and the hospital administration, bureaucratic procedures, patient cooperation, confrontation with death and dying, lack of staff support, and conflicts with physicians and other nurses (Gray-Toft & Anderson, 1981). Landsbergis (1988) found that among Swedish nursing home employees, burnout was significantly higher in jobs that combined high workload demands with low perceived control.

In addition to studying these stressors, our research assessed *self-esteem* because of its link to social comparison processes (Buunk, Collins, Van Yperen, Taylor, & Dakoff, 1990; Wills, 1991) and two other personality variables that may play an important role in the development of burnout. The first of these, *reactivity*, refers to a basic dimension of temperament which determines the intensity of reaction to both external and internal stimuli (Strelau, 1983). Specifically, highly reactive individuals exhibit stronger physiological stress reactions than less reactive individuals to an objectively identical stimulus. We included this variable not only because such individuals seem more susceptible to emotional exhaustion but also because there is some evidence that they are more inclined to give in to social pressure (Eliasz, 1980). Second, on the basis of recent developments within social exchange theory, we included the variable of *exchange orientation*, which refers to the personality disposition of individuals who are strongly oriented toward direct reciprocity, who expect immediate and comparable rewards when they have provided rewards for others, and who feel uncomfortable when they receive favors that they cannot reciprocate immediately. Such an orientation would make individuals in the human service professions more sensitive to the lack of reciprocity in the exchange with patients or clients (cf. Murstein, Cerreto, & MacDonald, 1977; Buunk & VanYperen, 1991).

The sample of our study was 351 Dutch nurses (Buunk, Schaufeli, & Ybema, 1992; VanYperen, Buunk, & Schaufeli, 1992). The sample (response rate 86%) includes members of different nursing disciplines and work settings: general nurses (37%), psychiatric nurses (14%), community nurses (8%), nurses working with the mentally retarded (21%), and nurses working in nursing homes or hospices (14%). The remaining 6% of nurses were employed in other health care settings. The nurses were mostly female (60%). Their mean age was 31 years (SD = 4.8) and they had considerable work experience in nursing (M = 10.4, SD = 4.5). Burnout was measured with the 22-item Maslach Burnout Inventory (Maslach & Jackson, 1986). According to a psychometric study of Schaufeli and Van Dierendonck (1992), the validity and reliability of the Dutch version is comparable to that of the original inventory. In this study, the internal consistencies (Crohnbach's α) of the three subscales were satisfactory: emotional exhaustion (α = .89), depersonalization (α = .71), and reduced personal accomplishment (α = .80).

DIFFERENTIAL CORRELATES OF THE BURNOUT DIMENSIONS

Our research indicates that although the three dimensions of burnout may have common roots, there are also quite different processes related to these dimensions (cf. Buunk, Schaufeli, & Ybema, 1990). An imbalance between investments and outcomes is directly related to all three burnout dimensions, and uncertainty plays at least some role with respect to all three aspects. The relationship of the burnout dimensions with the third stressor, lack of control, is somewhat weaker. Most interestingly, emotional exhaustion, depersonalization, and lack of personal accomplishment are each in part predicted by a unique interaction effect.

For emotional exhaustion, the interaction is between reactivity and uncertainty. As Figure 4-1 shows, when uncertainty is low, individuals high and low in reactivity do not differ much in degree of emotional exhaustion. Reactivity becomes especially important when uncertainty is high: emotional exhaustion is particularly found among highly reactive individuals who experience a high degree of uncertainty about how to feel and react. Apparently, a low level of reactivity buffers the negative effect of uncertainty on emotional exhaustion. Thus, less reactive nurses cope better with uncertainty in their jobs than their highly reactive colleagues. This interpretation is in line with Strelau's (1983) finding that less reactive individuals generally employ more effective active coping strategies, whereas highly reactive individuals are characterized by a less effective and passive coping style.

A very different pattern is found for depersonalization. First, this aspect of burnout is the only one that is directly related to self-esteem, and is clearly more characteristic of individuals with low levels of self-esteem. More interesting and theoretically important, however, is the apparently crucial role played by exchange orientation in combination with uncertainty. As Figure 4-2 shows, when uncertainty is low, individuals high and low in exchange orientation hardly differ in their level of depersonalization. But when there is a high degree of uncertainty, individuals high in exchange orientation respond with a much stronger tendency to devalue their patients than individuals low in exchange orientation.

VanYperen et al. (1992) examined the role of a related individual difference variable, communal orientation. This concept refers to the desire to give and receive benefits in response to the needs of and out of concern for others,

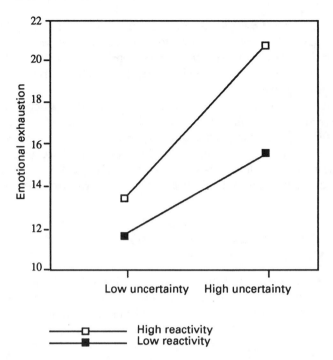

FIGURE 4-1 Emotional exhaustion: uncertainty x reactivity interaction effect.

and to help others when they are distressed (Clark, Ouellette, Powell, & Milberg, 1987). VanYperen et al. (1992) showed that for nurses high in communal orientation, imbalance in the relationships hardly mattered, but for individuals low in communal orientation, such imbalance was clearly related to burnout. Apparently, nurses who have no strong desire for reciprocity in their relationships with patients and who are responsive to patients' needs are not in danger of developing feelings of depersonalization. In the same vein, Cherniss and Krantz (1983) found remarkably little burnout in what they called "ideological communities." Workers in these settings (residential programs for mentally retarded people operated by a Catholic religious order) were strongly committed to the institute's patient-oriented ideology that was strongly "communal" in nature (see Chapter 8).

Personal accomplishment also has its own characteristic interaction effect that is theoretically meaningful. As is apparent from Figure 4-3, individuals with low self-esteem have a generally higher level of reduced personal accomplishment, i.e., they experience stronger feelings of inefficacy and demotivation. However, when they perceive a lack of control, individuals with high self-esteem experience nearly the same degree of lack of personal accomplishment as those with low self-esteem. Apparently, a lack of control is more important for individuals high in self-esteem than for individuals low in self-esteem. High self-esteem individuals may expect to be in control and feel their self-efficacy threatened when facing a work environment that is beyond their control (see Cherniss in Chapter 8).

In sum, we have come to view emotional exhaustion as an obvious indicator of general job stress that is more common among individuals sensitive to stress in general; depersonalization as a way of coping with the problems arising out of the relationships with clients; and lack of personal accomplishment as a response reflecting a low self-esteem and lack of control over the situation. We will now turn to the role of social comparison processes with respect to burnout.

SOCIAL COMPARISON, AFFILIATION, AND ISOLATION UNDER STRESS

Although it was emphasized above that stress leads to affiliation with, and information seeking about, similar others, research in this area offers a theoretical puzzle. On the one hand, Schachter and others have clearly shown that

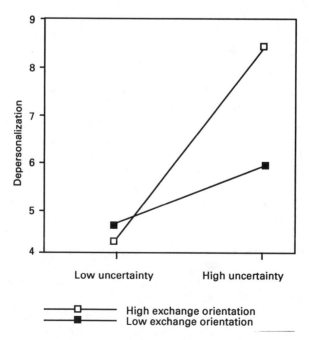

FIGURE 4-2 Depersonalization: uncertainty x exchange orientation interaction effect.

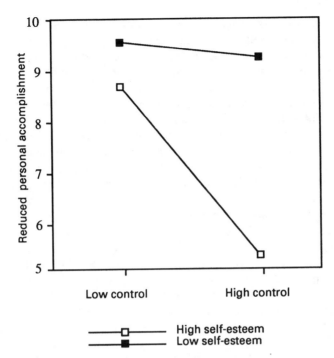

FIGURE   4-3 Reduced personal accomplishment; lack of control x self-esteem interaction effect.

stress, and particularly uncertainty about one's own reactions, leads to a tendency to affiliate with similar others (Wheeler, 1974). Indeed, as Pines and Aronson (1988) suggest, the sharing and testing of social reality is a basic function of social support systems at work. On the other hand, there exists experimental evidence that the tendency to affiliate is decreased in many stressful situations, particularly when one is confronted with embarrassing circumstances, or when the others are viewed as competitors (Sarnoff & Zimbardo, 1961; Teichmann, 1987). Talking with similar others about one's problems may be felt as admitting inferiority and may therefore induce a negative comparison that individuals try to avoid (Nadler, 1991). Especially within work organizations, persons may be concerned with maintaining an image of competence and may be reluctant to be among others when they themselves are confronted with negative emotions, particularly when these might be interpreted as indications of incompetence.

We reasoned that both viewpoints may be compatible, by assuming that on the one hand stress at work may foster the *desire* for information about and affiliation with others, but that on the other hand such stress may decrease *actual* affiliation with colleagues out of fear of embarrassment and of looking incompetent. This line of reasoning agrees with the utility affiliation theory of Rofé (1984), who asserts that the tendency to affiliate under stress depends on the perceived benefits and disadvantages of the company of others.

With respect to these issues, a number of important findings were obtained (Buunk et al., 1992). First, of the three dimensions of burnout, only emotional exhaustion was related to the desire for information about similar others and to the desire for affiliation, i.e., the desire to talk with others about problems at work. This validates the early work of Schachter (1959) in a somewhat different setting as emotional exhaustion seems, more than the other burnout dimensions, an indicator of general job stress. Second, as predicted by social comparison theory, it appeared clear that of the three stressors only uncertainty, and not lack of control or imbalance, was related to the desire for information and affiliation. Third, the desire to obtain social comparison information was much more obvious among those of high self-esteem, who seem especially inclined to respond to stress by seeking out information about similar others as well as the company of others. As expected, both uncertainty and emotional exhaustion were correlated with the actual avoidance of coworkers. However, this avoidance was also related to the other stressors and dimensions of burnout, providing strong evidence for the tendency to avoid others when under stress. Again self-esteem played an important role: avoidance was particularly apparent among those with low self-esteem who felt little control over their situation. For those high in self-esteem, a lack of control appeared to matter less.

These findings seem to indicate that those under stress—and particularly those low in self-esteem—respond to stress by avoiding their coworkers. One reason for this may be "pluralistic ignorance": the situation in which virtually all members of a group feel deviant and think their experiences are different from those of the others (Miller & McFarland, 1991). Thus, a nurse experiencing uncertainty may think that he or she is the only one with this problem and may therefore refrain from talking to others about it, while at the same time feeling the desire to do so. As Maslach (1982b) suggested, this is especially likely to occur because human service professionals tend to avoid revealing any personal thoughts or feelings that would be considered "unprofessional" and to act as if one were in control of the job and doing well.

UPWARD AFFILIATION

The next question is, with whom does one want to affiliate or about whom does one want to obtain information? As noted before, the Taylor and Lobel (1989) model predicts that under stress individuals will affiliate and seek information upward, that is with others who are doing better, for reasons of self-improvement. In contrast, Wills's (1981) downward comparison theory would predict that in such a case the company of others doing worse is preferred as a way of self-enhancement. There is, however, a third hypothesis that would emphasize the motive of self-evaluation. According to Schachter's original hypothesis, in a stressful situation people prefer to be with similar others, as this provides the best possibility for self-evaluation. It is senseless to compare one's feelings with others who differ too much in competence and experience.

Our study showed that there was a general preference for upward information in terms of competence and experience (Buunk et al., 1992). About half of the subjects expressed a preference for information about others who were more competent in their work. Of the others, most indicated a preference for others equally competent as oneself, and only a few were interested in information about others less competent. The upward trend was considerably more pronounced with respect to experience. About three out of four subjects preferred information about others more experienced, a minority were interested in others with the same level of experience, and virtually nobody was interested in others with less experience. We found some evidence that individuals faced with stress had a different information preference than individuals who were not under stress. The preference for upward information was characteristic of those low in reduced personal accomplishment, i.e., those feeling good about their own accomplishments at work. In contrast, nurses feeling bad about their accomplishments preferred information about others at the same level rather that at a more successful level. Contrary to the Taylor and Lobel (1989) model, a desire for upward information was not typical for individuals under stress.

There was also a general preference for upward affiliation. About half of the subjects expressed a preference for affiliation with others who were more competent in their work. The remainder all indicated a preference for others equally competent, and nobody was interested in affiliation with others less competent. The upward trend was considerably more pronounced with respect to experience. About three of four subjects preferred information about others more experienced, a minority were interested in others with the same level of experience, and virtually nobody was interested in others with less experience. However, the desire for upward affiliation was not at all related to stress. Thus, although upward affiliation was found among some subjects experiencing stress, it was not a unique characteristic of this group.

The desire for information was more upward than that for affiliation in the case of competence as well as in the case of experience. Apparently, despite the preference for information about more competent and experienced others, individuals are somewhat reluctant to obtain such information by seeking out the company of others who are doing better. These results are in line with findings from experimental research. For example, Smith and Insko (1987) found that subjects more often preferred information about the highest scoring other when they could examine such information alone than when they were made to believe that they would have to discuss the test with the chosen other. As Brickman and Bulman (1977) documented, individuals prefer to avoid exchanging comparison information under conditions in which they are obviously the inferior partner.

The data just presented offer no support for a number of assumptions of Wills's (1981) downward comparison theory. Few individuals, whether under stress or not, seemed interested in discussing their problems at work with less competent and less experienced others, or in obtaining information about such others. Additional data from our study, namely those on cognitive comparisons (Buunk et al., 1992), are also relevant with respect to Wills's theory. A majority of nearly 75% of the sample felt they were coping as well as most others in a similar situation. Of the remaining subjects, most felt they were coping better than others. Considering a second comparison dimension—how well off one was in general—there was an even stronger tendency to perceive oneself as better off rather than as worse off in comparison to similar others. No less than half of the sample felt they were better off than others, and only 13% felt they were worse off. About 40% felt they were as well off as the average nurse. From the perspective of Wills's (1981) theory it is important that, as far as there was a difference between those under stress and those who were not, the first group emphasized they were better off. Thus, in contrast to what Wills (1981) assumed, these data suggest that individuals under stress are not inclined to engage in downward comparisons by emphasizing that there are others who are worse off.

AFFECTIVE CONSEQUENCES OF DOWNWARD COMPARISON

Is there no truth at all in Wills's (1981) assertion that downward comparisons are engaged in as a way of coping with stress? Preliminary data from our research show that there is, although not in the way Wills thought. What we refer to here is that individuals under stress may derive positive feelings from seeing others do worse. There is some evidence that downward comparisons, i.e., information about others who are doing worse, may improve the mood of depressed and low–self-esteem individuals (Gibbons & Gerrard, 1991). We found that most of our subjects indicated that they felt bad when seeing colleagues doing worse than they were themselves, and only a minority reported positive affect in such a situation. Interestingly, however, this last type of affective outcome was characteristic for individuals under stress. All the stressors examined—lack of control, uncertainty, and imbalance—and all the burnout dimensions were related to deriving positive affect from downward comparisons. This suggests that, in line with Wills's (1981) theory, making downward comparisons plays a role in the coping process: nurses under stress engage more often in downward comparisons that made them feel better about themselves (Buunk, Schaufeli, et al., 1990).

DO SOCIAL COMPARISONS MATTER?

One of our assumptions is that social comparison processes within work units may contribute to the development of burnout. There is indeed some evidence from other studies that burnout is more likely to occur within certain work units than in others. For example, Edelwich and Brodsky (1980, p. 25) suggested that burnout in human services is like "staph infection in hospitals: it gets around. It spreads from clients to staff, from one staff member to another, and from staff back to clients. Perhaps it should be called staff infection." In a similar vein, Golembiewski and Munzenrider (1988, pp. 156–164) describe two large scale studies showing that between 70% and 86% of the employees classified as burned out were in work groups having at least 50% of their membership in the most extreme phases of burnout. They conclude that work groups have a tendency to develop homogeneous levels of burnout.

There may be several mechanisms through which social comparison processes within work units foster the development of burnout. First, research on "group polarization" would suggest that individuals who engage in a discussion of serious work problems in a group in which a negative view on these problems predominates will develop a more negative view after the discussion (Moscovici, 1985; Buunk, 1990) and may thus develop burnout symptoms. Second, colleagues may also directly influence the development of burnout symptoms by acting as models, whose symptoms are then imitated. In a process of "emotional contagion," individuals under stress may perceive symptoms of burnout in their colleagues and take on these symptoms, reasoning that these symptoms are apparently normal given their job situation (cf. Hatfield, Cacioppo, & Rapson, 1992; Schachter & Singer, 1962). Skelton and Pennebaker (1982) suggested that persons under stress may develop hypotheses about having a certain somatic disease, begin looking inside themselves for symptoms, and thus, in a process of self-fulfilling hypotheses, develop the symptoms associated with that disease. There is no reason to assume why a similar process could not occur with regard to burnout.

Given the fact that in our research those under stress tended to avoid others, it is interesting to note that Sullins (1991) showed in a recent experimental study that social comparison (observing another individual about to undergo the same experiment) induced mood convergence even in the absence of verbal communication. Moreover, negative moods appeared to be more contagious than positive moods, suggesting that burnout symptoms would be likely candidates for emotional contagion. Indeed, some evidence exists for such contagion as a precursor to burnout among human service workers (Miller, Stiff, & Ellis, 1988).

Social comparison theory would predict that not all individuals would be affected to the same degree by the symptoms they perceive in others. Those with a strong need for social comparison should be especially sensitive to the perception of burnout symptoms in others—and that is precisely what the data from our study of nurses suggest (Groenestÿn, Buunk, & Schaufeli, 1992). As Figure 4-4 shows, nurses who had a need to learn more about others in a similar situation expressed a higher level of emotional exhaustion when they perceived that many of their colleagues showed burnout symptoms.

Of course, it must be noted that these findings do not prove that among nurses with a high need for social comparison burnout is caused by the perception of burnout in others. It is also possible that these nurses project their burnout onto others as a way of validating the fact that they are burned out themselves (cf. Miller, Gross, & Holtz, 1991). Whatever the precise mechanism, the perception that others have similar symptoms may act to support the persistence of one's own burnout symptoms.

CONCLUSION

We have offered some preliminary evidence about the role of social exchange and social comparison processes in the development of burnout. It must be noted that thus far our research is only cross-sectional, that some of our findings on the effects of social comparisons are only suggestive, and that

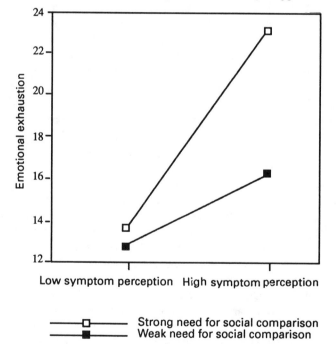

FIGURE   4-4 Emotional exhaustion: symptom perception x need for social comparison interaction effect.

longitudinal research is necessary to determine the exact nature of the causal links between the variables we examined. Nevertheless, our findings suggest that the social context plays an important role in the development of burnout. First, nurses facing uncertainty and emotional exhaustion feel a need to engage in social comparison and affiliation, particularly when they have high self-esteem. Second, nurses under stress seem reluctant to actually seek out the company of their colleagues, probably out of fear of looking incompetent, and such avoidance is particularly apparent among those with low self-esteem who feel little control over their situation. Third, nurses tend to prefer information about, and contact with, others who are more experienced and more competent. However, this tendency is more pronounced in the case of information seeking than in the case of affiliation, and it is not more manifest among those under stress. Fourth, in contrast to Wills's (1981) prediction, the feeling of being better off in general and coping better than others is more characteristic for those not under stress. Nevertheless, some nurses appear to deal with their stress by deriving positive feelings from downward comparisons, i.e., from seeing others doing worse. Fifth, there is some circumstantial evidence that nurses develop or sustain burnout symptoms because they perceive such symptoms in their colleagues, especially when they have a high need for social comparison.

Another conclusion that can be drawn from our results is that the three dimensions of burnout, though intercorrelated, reflect distinct psychological processes, as is evident from their somewhat different relationships with other variables. To put it simply, emotional exhaustion develops among highly reactive nurses who experience a high degree of uncertainty; depersonalization seems a way of dealing with uncertainty among those high in exchange orientation and low in communal orientation; and lack of personal accomplishment seems, although more characteristic of those with low self-esteem, a typical reaction for those with high self-esteem who are confronted with a lack of control.

Finally, our results indicate the importance of personality characteristics. For instance, self-esteem, reactivity, and exchange orientation moderate the relationship between stressors and particular burnout dimensions in nursing. Moreover, the nurses' level of self-esteem appears to be crucial for understanding social affiliation and social comparison processes in this particular profession.

Despite the provisional nature of our data, the present work has a number of implications for understanding and studying burnout in the human service professions. Most important, our findings show that the social nature of burnout has two aspects. First of all, professional burnout is related to an imbalanced social relationship with recipients. Burned-out nurses feel that they invest more in their patients than they get back in return. This agrees with the traditional view on burnout in which demanding and stressful relationships with recipients are considered to be the major cause of burnout (cf. Maslach, 1982b). We believe that social exchange theory can provide a useful conceptual framework to understand the dynamics of this particular social relationship that is critical for the development of burnout. However, the social nature of burnout is also demonstrated by the impact of social comparison and social affiliation processes in which coworkers are involved. In particular, we found indications for a process similar to emotional or symptom contagion in nurses. Accordingly, the traditional social psychological view of burnout that builds heavily on the demanding relationship between caregiver and recipient should be supplemented by a social comparison perspective that includes the relationships with coworkers as well. In doing so the gap between burnout research and social psychological theory can be partly closed.

Our results showed that burned-out nurses, particularly when they are under stress, do not affiliate with their colleagues; rather, they withdraw and avoid their presence. Viewed from a slightly different perspective, this agrees with the common notion that burnout is associated with a lack of social support from colleagues (Constable & Russell, 1986; Dignam & West, 1988). Our data suggest a plausible—albeit tentative—explanation for this well-documented relationship. Nurses under stress felt a strong need to affiliate, but at the same time they refrained from actually doing so. Therefore it is likely that their withdrawal is motivated by fear of embarrassment or of looking incompetent. Thus, it can be hypothesized that the lack of social support is caused by an active withdrawal from social contacts with colleagues who may confront the burned-out professional with his or her own incompetence. This interpretation is strengthened by the observation that the most vulnerable nurses with low self-esteem were especially likely to avoid the company of their colleagues. Furthermore, in an interesting study investigating both work and personal contacts (Leiter, 1988a), work contacts were positively related, and personal contacts were negatively related to emotional exhaustion. Thus, it seems that interactions with colleagues about work-related issues are stressful, but interactions about personal matters reduce stress. In conclusion, our findings suggest a counteractive role of social support from colleagues because work-related interactions might threaten the individual's self-esteem and foster burnout (cf. Buunk, 1990).

Finally, our study indicates that burnout has to be considered as a multidimensional phenomenon. Emotional exhaustion is a generic stress reaction that depends on individual sensitivity. We agree with Shirom (1989) that emotional exhaustion can be considered the core symptom of burnout. In our study too, emotional exhaustion showed the most robust and unambiguous results. Nevertheless, this aspect of burnout is the least specific since it overlaps considerably with similar strains (Schaufeli & Van Dierendonck, 1992). Depersonalization is a way of coping that is particularly prominent in individuals who are characterized by a strong need for reciprocity in social relationships. This second dimension of burnout, which is characterized by mental distancing, is highly specific for human service professions. Reduced personal accomplishment is observed especially in individuals with poor self-esteem. Probably personal accomplishment can be considered to be a coping resource that allows the individual to deal effectively with feelings of exhaustion, as is suggested by Koeske and Koeske (1989). Furthermore, Leiter (Chapter 14) confirms this special status of diminished personal accomplishment in showing that it develops separately from emotional exhaustion and depersonalization because the former is related to particular aspects of the work environment, such as lack of autonomy.

These observations concerning the three dimensions of burnout lead to two conclusions with regard to future research. First, the moderating role of personality characteristics should be investigated more thoroughly. Earlier studies on personality features such as self-esteem and locus of control (Caron, Corcoran & Simcoe, 1983) and "hardiness" (McCranie, Lambert, & Lambert, 1987) did not investigate such moderator effects on burnout at all. Second, comprehensive models should be developed that integrate all three dimensions of burnout. In our view a social comparison perspective can provide an overarching conceptual framework to develop such models.

TABLE 1 Burnout Dimensions, Stressors, and Personality Characteristics

	R^2	β
Emotional exhaustion	.36	
Uncertainty		.50**
Imbalance		.25**
Uncertainty × reactivity		.22**
Lack of control		.20**
Depersonalization	.32	
Uncertainty × exchange orientation		.30**
Imbalance		.24**
Self-esteem		.23**
Reduced personal accomplishment	.36	
Uncertainty		.24**
Imbalance		.25**
Lack of control × self-esteem		.25**
Reactivity		.11*

*$p < .05$. **$p < .01$.

II

INDIVIDUAL APPROACHES

Individual approaches have focused on the symptoms displayed by burned-out individuals. Such approaches were especially popular during the pioneer phase of burnout work. Burnout was considered a syndrome consisting of many related symptoms, of which exhaustion was the most prominent one, and long laundry lists of individual burnout symptoms were drafted. Frustrated expectations and goals were regarded as the major cause for burnout. It was observed that idealistic and strongly motivated individuals, who are extremely dedicated to their work and who are overinvolved in their jobs, were more likely to burn out. Although some authors emphasized the process of burning out, the prevailing view was a more static one that concentrated on the end state: the burnout syndrome.

The first two contributions criticize this state conception of burnout in similar ways. According to both Burisch and Hallsten, the state conception of burnout is overinclusive. Too many symptoms are associated with it, so that burnout cannot be discriminated from other mental states such as stress, depression, and alienation. Moreover, Burisch argues that the burnout literature has ignored many relevant research traditions (e.g., crisis theory, incentive theory, psychosomatics), all of which have dealt with something like burnout from various perspectives. Although Burisch and Hallsten share most criticisms on the present burnout conception, they offer different alternatives. However, both authors emphasize that burnout should be studied as a process rather than as a state.

Burisch maintains that burnout is best conceptualized as a fuzzy set, i.e., a highly nonspecific entity, the generic name for certain types of crises that manifest themselves in a multifaceted symptomatology. He identifies loss of autonomy as the salient causal factor: failing to get what one wants or having to endure what one wants to avoid. Rather than hoping for a circumscribed cause of burnout to emerge (something akin to a burnout virus), investigators should study what is actually happening during the course of burnout. Burisch outlines an action model that employs an action episode as its basic unit of analysis. The individual's latent motives lie at the core of the action episodes. Burisch identifies four types of disturbed action episodes, each of which may eventually lead to burnout. By offering his action model, Burisch provides a framework for more sophisticated theory and research.

Hallsten proposes a quite different framework by arguing that burnout is essentially a special kind of depression that results from a gradual process of "burning out." Burning out is assumed to occur when an active, self-defined role is threatened. Drawing on various theoretical frameworks (e.g. reactance theory, learned helplessness, and ego psychology), Hallsten identifies some key factors that contribute to burning out: personal vulnerability, strength of goal orientation, and perceived environmental incongruence. Depending on different combinations of these factors, different outcomes can occur. For instance, persons characterized by high personal vulnerability and strong goal orientation who do perceive environmental incongruence are expected to burn out. Other combinations lead to "alienation," "balanced commitment," or "circumscribed frustration." Thus, Hallsten suggests a complex etiological model that describes the process of burning out as well as processes leading to more positive outcomes. His analysis (much like that of Burisch) presents a valuable differentiation of the overinclusive state concept of burnout.

In the final contribution to this part of the book, Hobfoll and Freedy apply the overarching framework of the conservation of resources (COR) theory to burnout. COR theory is a basic motivation theory that postulates that stress occurs when resources are threatened, when resources are lost, or when individuals invest resources without the expected payoff. Resources are defined broadly as valued objects, conditions, personal characteristics, or energies. According to Hobfoll and Freedy, burnout is more likely to occur when resources are lost than when resources are not gained. They call this "the primacy of loss" and the "secondary importance of gain," respectively. Moreover, when loss occurs, or when resources are threatened, people engage in active coping in order to regain resources or to prevent their loss. From this perspective, burnout can be considered as the ultimate price that has to be paid for this kind of coping. In discussing burnout interventions from a COR perspective, Hobfoll and Freedy argue that greater emphasis should be placed on obtaining objective resources than on changing perceptions and cognitions.

None of these three individual approaches to burnout has yet been fully tested empirically, although the authors present indirect empirical evidence. For the time being, these approaches, in which the individual's motivations and expectations play a major role, have to be considered as heuristics or vehicles of thought that might inspire future research.

5

IN SEARCH OF THEORY: SOME RUMINATIONS ON THE NATURE AND ETIOLOGY OF BURNOUT

Matthias Burisch
University of Hamburg

Reading the theoretical literature on the burnout syndrome for the past decade, one sometimes gets the impression that what is being looked for is a specific cause of a specific mental disorder. If it has a name, we will find what lies behind it—something akin to a metabolic deficiency or a virus. Could it be role conflict, lack of feedback, an external locus of control, or too much red tape? Correlations between these and burnout measures keep being significant, given adequate sample sizes, but not very impressive numerically. And so the search goes on.

In contrast, my main thesis here is that burnout, if conceptualized appropriately, has in fact been *over*explained. This is because in several fields of inquiry knowledge has accumulated which taken together gives burnout the status of a fairly well-established term. Well established and respectable, that is, if judged

This chapter has profited considerably from detailed comments by the editors, Christina Maslach and Wilmar Schaufeli, and also from discussions with Cary Cherniss, Daliah Etzion, and Ayala Pines. Many of its remaining weaknesses result from my stubborn refusal to heed their advice; I will have to put up with the consequences.

by the standard of other terms in "soft" psychology, such as clinical, personality, and organizational psychology. (*Neurosis* as a term is a good example for comparison. What used to be a household word is now beginning to disappear from diagnostic manuals.) So much for the good news. The bad news is that with burnout being a rather nonspecific phenomenon, the explanations offered by various theories are also bound to be very nonspecific. Before I deal with the latter, let me sketch out what I consider an adequate conceptualization of burnout.

BURNOUT AS A FUZZY SET

Even before tackling the issue of definition, I would like to make the point that it is a futile question to ask, "What *is* burnout?" Because burnout is a construct, residing in the heads of many people (and probably a little differently in each one of them), all we can meaningfully ask is, "How do we want to *define* burnout?" in order to make it a useful concept. George Kelly is reported to have made essentially the same point in a different context, during a round table discussion. He said: "When I say, 'Professor Cattell's right foot is an extravert,' you look at his foot. Why don't you look at me?" His point there was that the temptation to reify a newly created term is very strong. In our case, that temptation may even override the realization that writer A's burnout is most probably not identical to writer B's.

So what is meant by "burnout?" When we turn to the literature for answers, we find that defining burnout is like defining the exact boundaries of a large cloud. Given that state of affairs, the delineation of the syndrome and its separation from other entities such as depression are badly needed, but this endeavor is not likely to succeed easily.

In my view, burnout is used as a generic name for certain ill-defined types of crises. It is a fuzzy set of symptoms or a fuzzy set of people with symptoms. Both sets overlap considerably with neighboring sets. But what is a fuzzy set? In general, a set is a group of elements for which there is no doubt about their membership in that group. The set of even numbers consists of all numbers that can be divided by 2. The set of Scandinavians consists of all Swedes, Danes, Norwegians, Finns, and Icelanders. The set of four-letter words includes all words that have four letters. But what about flowers? Most would agree that a rose is a flower, but why not a blossoming chestnut tree? For the set of sports, if tennis is a sport, then what about chess? Or consider the set of tall people; where is the threshold? These are sets where membership is a matter of opinion, and thus "fuzzy"—some elements are very central to the set, whereas others are more peripheral and may belong to neighboring sets as well. My argument is that this is the case with the elements of burnout.

Anyone inclined to take issue with the assertion that burnout is a fuzzy set of people should try the following exercise. Imagine a real person of your acquaintance whom you consider to be a prototypical case of burnout. Now try to imagine a second person, preferably matched as closely as possible with the first one in all relevant aspects (most crucially in severity of suffering), but not a case of burnout. It is an amazingly hard task. Most of us, I believe, would agree not to subsume a hallucinating psychotic under the burnout label, nor would we include a patient with Alzheimer's disease or hepatitis. But I suspect that we would be hard-pressed to draw a sharp line for a large part of the thousands of clients who populate psychotherapists' offices because they feel less happy than they would like to be. I also presume that those clients would have elevated scores on standard measures of burnout. So much for the people fuzzy set.

As for the symptom fuzzy set, several years ago I compiled all the burnout symptoms cited in the literature then available to me. A few articles did provide their own symptom lists, and so I assembled everything I found into one long list. If it was not an exhaustive list, it very nearly was. After elimination of synonyms and some more pruning, I counted more than 130 symptoms that I grouped, somewhat arbitrarily, into 11 subordinate and 7 superordinate categories (Burisch, 1989, p. 12). However, none of these many symptoms is unique to the burnout syndrome, i.e., *not* to be found in other nosological entities, such as depression.

It is of course viable to operationalize burnout in terms of questionnaire scores. Quantitative research involving decent sample sizes will probably have to resort to that option. Applied to individual cases, this strategy runs into the problem of arbitrariness. Savicki and Cooley (1983) suggest that scores on all three scales of the Maslach Burnout Inventory (MBI) should be elevated before we speak of burnout. They wisely refrain, however, from offering any cutoffs. Pines and Aronson (1988, p. 218) do provide one such cutoff: a score above 4 on their Tedium Measure (TM) is said to indicate "burnout to the extent that it is mandatory that you do something about it." However, that is analogous to stating that people must stand higher than 180 cm in order to be called tall. There is certainly nothing special about a score of 4. Of Kleiber and Enzmann's (1986) 69 helping professionals, 26% scored above 4. In Frank's (1989) sample of 217 nursing students, 20% did so. In Bode's (1988) sample of 87 nurses, 16% did so. Moreover, the Tedium Measure's fairly unspecific item content raises doubts as to its discriminant validity (cf. Burisch, 1984b).

Thus, there seems as yet to be no satisfactory way of defining burnout, and progress toward understanding it is hampered by the fact that it is an undefined entity that is being discussed. Again, this state of affairs is not altogether atypical in psychological nosology. Burnout has a certain gestalt quality, including configurations of symptoms, lifestyles, modes of thinking, job situations, and so on. As was said above, those configurations (implicit as they usually are) are likely to differ a little for each perceiver of the gestalt.

All I can do here is try to explicate what constitutes the core of *my* concept of burnout symptomatology, granting at the outset that my concept is no better than that of anybody else.

People whom I consider to be in some stage of a prototypical burnout process have one or more (typically all) of the following characteristics to a degree. For the moment, let me suggest these as core symptoms, while admitting that the terms lack precision.

- Hyper- or hypoactivity
- Feelings of helplessness, depression, and exhaustion
- Inner unrest
- Reduced self-esteem and demoralization
- Deteriorating or deteriorated social relationships
- Some active striving to bring about a change (a characteristic that distinguishes burned-out individuals from people mourning some loss)

RESEARCH TRADITIONS RELEVANT FOR UNDERSTANDING BURNOUT

This loosely defined syndrome has been described and investigated in a number of research traditions. In fact, burnout researchers have rediscovered something that has long been known elsewhere by other names. Fields of research that contribute to an understanding of burnout include:

- Crisis theory
- Frustration and aggression
- Reactance and learned helplessness
- Incentive theory
- Psychosomatics
- Psychology of conflict

Of course, it is impossible to do justice here to all these vast literatures. Nevertheless, a few words on each of them is in order.

Crisis Theory

Crisis is a term almost as ill defined as burnout. However, one handy definition reads as follows: "A system is in a state of crisis when its repertoire of control mechanisms or its energy resources are insufficient to handle problems which threaten the system's survival or its capacities for future development" (Bühl, 1984, quoted after Ulich, 1987, p. 6).

If we are willing to adopt a liberal notion of survival, then it becomes apparent that burnout is indeed a crisis. People do not burn out when their psychic survival is unthreatened and when they feel free to develop in any direction they want. Rather, burnout processes often start when some goal or goals (tangible or lofty, but sufficiently important) have remained unfulfilled for a long enough time despite attempts to reach them.

Frustration and Aggression

In the case of an active goal, this condition is known as goal frustration, and research on its consequences dates back more than half a century (Dollard et al., 1939). The original hypothesis that frustration is a necessary and sufficient condition for aggression has proven untenable. However, there is some consensus (plus confirmation in plenty of everyday observations) that frustration is likely to arouse the hostile emotion of anger. Whether the anger is acted out and, if so, against whom depends on a host of situational factors (Yates, 1962).

Stokols' "psychological theory of alienation" deals with a more circumscribed frustration situation, but it could pass as a theory of burnout in people-oriented professions. It describes "a sequential-developmental process which (a) develops in the context of an ongoing relationship between an individual and another person or group of people; (b) involves an unexpected deterioration in the quality of the outcomes provided to the individual by the other(s); and (c) persists to the extent that the individual and the other(s) remain spatially or psychologically proximal" (Stokols, 1975, p. 26). This, of course, is precisely the situation of the beginning psychotherapist, nurse, drug counselor, or parole officer who needs some optimism to start working, but has to face continually frustrating and all too proximal clients. As to effects, Stokols says: "Recent research literature on social psychological stress . . . would suggest that when P is unable to alleviate his experience of alienation, and this experience extends over a prolonged period of time, two general syndromes of stress may ensue: psychological stress as reflected in certain physiological disorders . . . , and self-destructive or antisocial behavior as manifested in P's self-disparagement or aggression towards others" (p. 35). Here is one simple explanation for a number of core burnout symptoms (particularly the depersonalization component), for the family problems of many burned-out individuals, for their withdrawal from friends, and for their quarrels with colleagues—all of which I have lumped together under "deteriorating or deteriorated social relationships."

Reactance and Learned Helplessness

Much the same, plus some more, is found in theories of reactance and learned helplessness, although researchers from these latter two traditions rarely seem to have taken notice of the frustration-aggression school of thought, and vice versa. Reactance theory originally dealt with the situation where some freedom was being threatened or eliminated. Under these circumstances, individuals are hypothesized to become motivationally aroused to restore their freedom and will feel hostile and aggressive. These effects will be particularly strong if individuals started with the expectation of being free. Learned helplessness theory, in turn, covers the later stage when efforts at restoration of freedom have failed and the individual is giving up or has given up. The theory states that motivation to expend energy is then reduced and feelings of depression set in. Whereas the motivational effect is still being debated, the effect of helplessness on mood seems to be widely acknowledged. Wortman and Brehm's (1975) well-known integration of reactance and learned helplessness theory has a two-stage process: first hyperactivity, then hypoactivity, with accompanying emotional states. This is precisely what we see so often in people burning out.

Kahn, Wolfe, Quinn, Snoek, and Rosenthal (1964) devote a full chapter of their classic *Organizational Stress* to the built-in frustrations or helplessness experiences of "new-guard" members working in organizations dominated by an "old guard" stifling their efforts. The authors illustrate their discussion with the case of an innovator in "an organization [which] has ostensibly committed itself to scientific principles of self-improvement, but summarily rejects any suggestion of change" (Kahn et al., 1964, p. 131). Harry Levinson (1981, p. 76), the eminent management theorist, says that "a special phenomenon occurs after people expend a great deal of effort, intense to the point of exhaustion, often without visible results. People in these situations feel angry, helpless, trapped, and depleted: they are burned out."

Incentive Theory

Essentially the same drama is the subject of incentive theory, a general theory of motivation developed by Klinger (1975, 1977). In place of freedom or freedoms, his central concept is incentive, i.e., any object or event that attracts or repels an individual. The concept seems to have much in common with what Hobfoll (1989) calls "resource" (see also Chapter 7). What happens when an individual has formed a commitment to an incentive that turns out to be unattainable, thus forcing the individual to disengage from it? To quote Klinger (1975, p. 8): "Dissolving an important concern should tend to produce effects such as apathy, reduced instrumental striving, loss of concentration, and increased preoccupation with momentary cues, a pattern that is recognizably similar to depression." I would add that this pattern is even more similar to burnout, particularly when we remember that Klinger is speaking here of undeniable failure, whereas the incentives that motivate many burned-out individuals (such as proving oneself to be a competent counselor or the world's leading burnout researcher) are usually too nebulous to ever permit a clear-cut failure or goal attainment.

Klinger's theorizing is particular intriguing because its underpinnings are based on a great deal of empirical work, even animal studies. Nevertheless, his thinking is sufficiently freewheeling to encompass such concepts as existential meaning. In fact, one of his books is entitled *Meaning and Void* (Klinger, 1977), which links nicely with Pines' and Hallsten's view on burnout (see Chapters 3 and 6, respectively).

Psychosomatics

Psychosomatics is another field of relevance for burnout. Craig and Brown (1984) compared 135 patients with gastrointestinal disorders with a matched control group of healthy subjects. Fifty-six (or 41%) of the patients were diagnosed as organic cases. When the recent life histories of all participants were searched systematically, it was found that only 9% of the healthy controls had experienced a major goal frustration during the 38 weeks prior to the interview. In the subgroup of organic patients, that rate was 54%, or six times as high. Moreover, the authors report that the severity of the patients' frustration was typically exacerbated by their own actions, e.g., taking a hopeless case to an appeals court. Interpreting their findings vis-à-vis earlier dispositional hypotheses, Craig and Brown (1984, p. 416) write: "It seems likely that having ambition or working industriously toward desired goals is insufficient in itself to bring about organic illness. Our findings suggest by contrast that it is the *frustration* of these plans and ambitions that is of prime importance. In short, it may well be that there are many who strive earnestly, but that relatively few experience severe goal frustration as reported by us, and it is only these few who are at risk of becoming ill."

Ulcers are not the only pertinent entities here. Other examples include cardiovascular disease and cancer. To illustrate, let me quote from a review by McQueen and Siegrist (1982, p. 362) linking social factors and chronic disease:

Stressors which threaten socioemotional bonds and/or maintenance of social status can be responded to by active or passive coping. During active coping, the individual has the feeling that it is necessary and possible to fight against threats (predominance of the defense response), whereas passive coping can be characterized as a giving-up reaction after experiences of powerlessness and/or helplessness (predominance of the conservation-withdrawal reaction). Active coping without success, continuous struggle without reward, intense threat to one's efforts to control a relevant situation, and exorbitant or overwhelming demands upon one's adaptive capacities—these seem to be classes of critical experiences which create feelings of irritation, anger, frustration and dissatisfaction. It is probable that during these experiences both stress axes are activated, i.e., the sympathetic-adrenal-medullary and the pituitary-adrenal-cortical system. It will be suggested . . . that simultaneous activation of these two stress axes is important in creating neurohormonal imbalance which affects the cardiovascular system. On the other hand, passive coping combined with distressing emotional states of hopelessness and helplessness is thought to activate the pituitary adrenal-cortical system only, and to increase vulnerability to infectious processes and to malignant developments.

The authors add, cautiously, that "much of this working hypothesis . . . still waits for further confirmation" (pp. 362–363).

Psychology of Conflict

Finally, I was fascinated to find a German textbook of psychiatry which as early as 1969 described a so-called exhaustion reaction as follows (Bräutigam, 1969, p. 32):

Typically, there is a strained, irritable exhaustion which, unlike the comfortable weariness after success, includes a morose ill-humor and lowered capacities. The feelings of impotence and fatigue are accompanied by a state of tension. The exhaustion and weakness does not give way to recovery in peaceful sleep. Rather, there is usually a paradoxical inability to relax and sleep. Complaints of tiredness, slackness, and incapacity to achieve can thus be in the focus. However, in most cases this contrasts with a remarkable unrest. Mood is characteristically not outright depressive but shallow, stale, empty, listless. Everything is too much, everything is a demand one would rather not have any more. Attention is often directed to one's own body. A hypochondriacal, plaintive, grievous tendency with mostly diffuse complaints shows up.

In my opinion, this is as good a symptomatic description of burnout as any other. As to the etiology, the author has this to say (Bräutigam, 1969, p. 34):

The crucial factor seems to be a certain type of achievement-related conflict which can be characterized as "self-contradictory effort" (von Baeyer). The pathogenetic conflict of the exhaustion reaction lies in an approach-avoidance ambivalence concerning one's own efforts. The motivation to achieve is hampered by the obstacle of some inner contradiction, the awareness of its futility, or the like.

I have taken the reader on this trip through some neighboring fields of research to demonstrate that there is a wealth of knowledge that has gone largely unnoticed in the burnout literature. With the exception of learned helplessness, I have rarely if ever come across any explicit reference to the fields just listed. Admittedly, there may be subtle differences in the conceptualizations of symptoms and alleged causes by the various researchers I have quoted. However, given the coarse grain of the language we converse in and the elusiveness of the burnout construct, any attempts at more fine-grained analyses may be premature. True, we cannot be sure that Craig and Brown's goal frustrations refer to the same experiences as McQueen and Siegrist's powerlessness. But, to say the least, the corresponding doubt about whether Freudenberger and Maslach have the same sort of people in mind when they speak of burnout is by no means smaller.

A HIERARCHY OF THEORETICAL APPROACHES TO BURNOUT

Approaches to the study of burnout can be grouped according to their level of abstraction or generality. I find it convenient to distinguish four such levels (see Figure 5-1) and will comment on each one of them.

Top Level: Loss of Autonomy

The research traditions I outlined in the previous section have one common core: loss of autonomy as the critical factor. To me, this represents the most abstract level of explanation for burnout. To paraphrase Wortman and Brehm (1975, pp. 282–283), autonomy refers to the sense that one can do as one wants, that one does not have to do what one doesn't want, and that one does not have to endure what one wants to avoid. Obviously, this encompasses much more than, for example, decision latitude, a variable that can be measured directly and is sometimes also labeled autonomy in organizational research. I prefer the term autonomy to others, like freedom or control, because autonomy may even imply the freedom to give up control voluntarily (e.g., by becoming a monk or riding a roller-coaster).

For a burnout process to start, of course, either a very central aspect of autonomy must have been lost, or the loss must be very pervasive, i.e., generalized across many aspects. I would even go one step further and claim, in line with the quoted cause of the exhaustion reaction, that in burnout the crucial autonomy loss stems from an inner conflict, of either the approach-avoidance or the avoidance-avoidance variety. A police officer who stays on the job in hope of some future promotion, although he needs more and more alcohol to shake off his daily stress (cf. Maslach, 1982a, pp. 72–73), may provide an example of an approach-avoidance conflict leading into burnout. An avoidance-avoidance conflict would be exemplified by the free-clinic director who, although he has to push himself to work, carries on because he feels he cannot leave his clients alone (Freudenberger & Richelson, 1980, p. xviii).

In other words, burned-out people are either blocked or trapped. Insofar as those blocked find no way to circumvent the blockage or to let go, both types are trapped. I am willing to say this because it is what I have seen in every single case of burnout I have been involved with so far. I have to concede, of course, that this sort of clinical evidence is not considered hard data by common

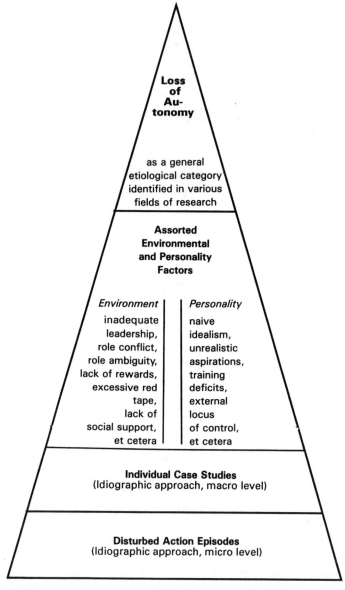

FIGURE   5-1 Levels of analysis for the Study of burnout

standards. Nevertheless, when we accept autonomy loss as the crucial causal factor in burnout, then its symptomatology becomes much more understandable. Most or all burnout symptoms can be interpreted as either direct effects of autonomy loss or as attempts to prevent further loss, to regain autonomy, to compensate for its loss, or to lessen its subjective impact.

Pines (Chapter 3) prefers to view "failure in the existential quest for meaning" as the "root cause of burnout." It is my experience, however, that people are most likely to question their own life's meaning, and the meaning of life in general, when confronting a permanent loss of autonomy. And indeed, Pines's burnout model contains the stages of "goals and expectations not achieved" and "failure," which are conceptually fairly close to autonomy loss, as prerequisites for burnout. Thus, she focuses on a later link in the causal chain than I do. As I see it, failure to achieve a sense of existential significance (and suffering from it) is characteristic of advanced levels of burnout, the terminal stages of a process triggered by some loss of autonomy.

Klinger (1975) argued convincingly that what he calls the incentive-disengagement cycle (namely, an orderly sequence of invigoration, aggression, depression, and eventual recovery) is a biologically based, unlearned response to incentive loss or, as I call it, autonomy loss. That would explain the hyperactivity so often seen in the early stages of a burnout process as well as the later hypoactivity, the lack of energy, the feelings of depression and exhaustion, the reduced self-esteem, the continual inner unrest, and the deteriorating relationships with part or all of the social environment. The withdrawal from work or clients may serve the function of making further autonomy loss less likely. The dehumanizing attitudes and distancing behaviors, in addition to being outlets for aggression, may be attempts to regain autonomy. The self-preoccupation of many burned-out people, their increased drug use or food consumption, their absenteeism and "chrysalis" behavior (Lauderdale, 1982) may compensate for some autonomy loss and lessen its subjective impact.

Second Level: Environmental Factors

"Failing to get what one wants" or "having to endure what one wants to avoid" denotes a very unspecific class of experiences. Therefore, researchers around the world have begun to look for the presence of negative features (such as role conflict, role ambiguity, or too much red tape) and the absence of positive features (such as feedback, recognition, or adequate pay) in the environments of people who are burning or burned out (see Chapter 3).

Regrettably, the predominant strategy so far has been to correlate some questionnaire measure of burnout with some concurrent questionnaire measure of the targeted factors. Roughly, this amounts to asking, "Do you feel unhappy about yourself?" and "Do you feel unhappy about the things around you?" and correlating the answers to these questions. Given the high degree of content overlap and the probable influence of response styles (plus the likelihood that people feeling bad before entering their jobs will later tend to describe their surroundings more negatively), the magnitude of the correlations emerging from this type of design is surprisingly low. For instance, in Enzmann and Kleiber's (1989) study of 130 helping professionals, the best predictor of any burnout scale was a six-item scale reflecting the feeling of being "overtaxed" on the job. This scale correlated at $r = .60$ with the MBI's emotional exhaustion component and $r = .67$ with Pines's TM. All of the items, however, could have been included in any standard burnout inventory on the basis of their content, which somewhat reduces the surprise value of these findings. Even with this predictor in two of the equations, multiple correlations between job (and some personality) factors and burnout were only $R = .72$ (MBI emotional exhaustion; four predictors), .46 (MBI personal accomplishment; three predictors), and .34 (MBI depersonalization; two predictors), figures that may be expected to shrink a little under cross-validation.

Likewise, the correlations between position in Golembiewski's MBI-based eight burnout "phases" and either organizational features or consequences of burnout are often far from perfect in his own data (Golembiewski, Munzenrider, & Stevenson, 1986). These correlations were even lower in several unpublished studies done at our institute. In one of these (Frank, 1989), all 227 female students in a large nursing school were given a modified version of the MBI (both the frequency and the intensity scale were replaced by a single 7-point agree–disagree response scale) and an almost unaltered version of the German edition of the Tedium Measure (Aronson, Pines, & Kafry, 1983). The students were also given a questionnaire covering various work setting factors regarding their school and the wards where they received practical training. Those factors included, for example, quality of supervision and general work climate. Lacking norms from a sufficiently large and representative sample, we used within-sample medians as cutoffs to assign individuals to one of the eight Golembiewski categories. Moreover, unlike Golembiewski, we left the MBI scoring key unchanged. When mean ratings for the work setting factors were correlated with the rank order position in Golembiewski's eight-phase model, τ coefficients ranged from $-.21$ to .91, averaging .51.

In another study (Bode, 1988), essentially the same procedure was followed with a stratified random sample of 87 practicing nurses. Here the τ coefficients ranged from .00 to .71, with an average of .47. Finally, Moritz (1991) gave the MBI and the TM to 216 employees (73% of whom were male) of an industrial corporation, as part of a general climate survey. The τ coefficients for the questionnaire's six scales ranged from $-.14$ to .86, averaging .29. For comparison, the book of Golembiewski et al. (1986) contains a total of 68 variables, both worksite factors and personality or symptom scales, where phase means can be correlated with phase rank order position. The resulting τ coefficients vary from a low of .07 to a high of 1.00, with an average of .61.

Remember that those are correlations between subsample means and a weighted linear combination of dichotomized MBI scores. The corresponding correlations across individuals would be even lower.

Keeping the well-known shortcomings of cross-sectional designs in mind, what about longitudinal studies? Cherniss (1980a), in his now classic investigation of 28 "new public professionals," compared the eight "most burned out" and the six "least burned out" of his respondents and found very high correlations between that dichotomy and eight work setting factors, also dichotomized. These factors, highly correlated among each other, included quality of the orientation process; workload; intellectual stimulation, challenge, and variety; scope of client contact; amount of professional autonomy; clarity and consistency of institutional goals; quality of supervision; and degree of social isolation on the job. Calculations of the G index of agreement (Holley & Guilford, 1964) for Cherniss's data result in figures between .57 and .86, with an average of .79. In contrast, corresponding figures in Frank's (1989) study of nursing students ranged from −.06 to .40, with an average of .25, and in Bode's (1988) study of nurses from .11 to .33, with an average of .22. Our samples were homogeneous with respect to type of job whereas Cherniss's sample comprised four dissimilar occupations, which may explain the divergent findings.

Jackson, Schwab, & Schuler (1986), in a comparatively well-controlled study, used job factor ratings from 248 school teachers to predict MBI scores a year later. Although several of their specific hypotheses found support at a statistically significant level, the authors concede with refreshing frankness "the generally low amount of total variance explained by all job conditions assessed in this study" and begin to consider "the possibility that our hypotheses about the antecedents of burnout are inadequate" (Jackson et al., 1986, p. 639).

Second Level: Personality Dispositions

Could the lack of impressive covariation between environmental factors and burnout be due to neglect of a crucial class of variables—the personality backgrounds of those who seek out, shape, perceive, and react to their environments? After all, even the most burnout-breeding school system, hospital ward, or sales department usually has some individuals who do *not* burn out, although they seem not to differ conspicuously from others who do. Autonomy loss is such an ubiquitous experience that all of us would have to be burning out all the time if that were all there were to it. But this does not seem to be the case. Loss of autonomy is necessary, but probably not sufficient to trigger a burnout process. It must meet an individual with certain dispositions.

A lot has been written about what exactly these dispositions are, and a host of constructs have been invoked to fill that theoretical gap. Unfortunately, because ethical considerations practically rule out experimentation in burnout research, only longitudinal studies have a chance to shed some light here. More specifically, prospective studies are needed to assess personality characteristics in large samples of subjects *before* some critical point in their lives, e.g., before starting a career. None of the few longitudinal investigations I know about have employed such a design.

But even from such methodological refinements we should not expect too much. In line with a pessimistic position sketched by Jackson et al. (1986, p. 637), I am afraid that use of conventional research paradigms for another 20 years will result in more data but not much more knowledge. Even more sophisticated investigations of burnout etiology resemble an attempt to predict the outcome of a football match on the basis of the teams' average height and weight on the one hand, and from the prevailing wind direction and speed on the other. What is more, outcome in terms of scores is probably not the most interesting aspect of the game.

Third Level: Individual Case Studies

One of the benefits of doing burnout research is that one is entitled to read novels and biographies during work hours. In fact, Graham Greene's 1960 novel *A Burnt-out Case* used the term more than a decade before Freudenberger's seminal 1974 article stirred up tremendous interest. My favorite burnout character from the world of fiction, Senator Thomas Buddenbrook, is one of the heroes in Thomas Mann's *Buddenbrooks* (first published in 1900); he dies in 1875. Biographies of luminaries such as Goethe (see Ipser, 1987), Hesse (see Freedman, 1978), or Wittgenstein (see Bartley, 1983) contain accounts of burnout-like crises, some passing, some long-lasting, some lifelong. Without doubt, there must be many more examples.

The professional literature on burnout is replete with real-life vignettes. Yet more detailed case studies, running to some length and supplying more than the biographical highlights for which the particular writer wanted an illustration, are notably lacking. So I can be very brief here. Published case collections could go a long way toward delineation of the syndrome. I have grave doubts that a panel of experts, independently diagnosing a sample of individuals as burning/burnt out or not, would produce anything near unanimity. As long as we are not sure we are talking about the same phenomenon (i.e., the same fuzzy set), progress in the way of meaningful and replicable results is unlikely to occur. This is an important issue that needs clarification. Moreover, case histories that cover some span of time and describe interventions, as well as their outcomes, could provide clues as to what works and what does not in attempts to ameliorate the lot of burned-out people.

Fourth Level: Individual Action Episodes

At the most concrete, or molecular, level of analysis we find single action episodes, and this is the level in which I am currently most interested. My advocacy of an idiographic approach at this fine-grained level stems from the conviction that in the study of a phenomenon such as burnout, one ought to move to the higher levels, and to data aggregation across individuals, only after a sound understanding has been gained of what is actually going on in the life of each of them. True, this is not typically the level at which burned-out individuals themselves prefer to think. However, it can be a most enlightening (and therapeutic) exercise to confront them with the realities of what they have experienced. For instance, I have heard school teachers complain that their work is made unbearable by "discipline problems" in their classes or that they "just don't get along with students any more." But further probing revealed that they had overgeneralized from two or three out of several dozen students they were teaching and that they were referring to very few critical encounters in which they had been unable to assert themselves. What to an unconcerned outsider would have looked like fairly innocent incidents had taken on a particularly ego-bruising significance to these teachers.

Support for this level of analysis is found in the writings of Eugene Heimler (1975, 1985), a British social worker who for decades has consulted with the unemployed and demoralized individuals in general. Among his methods is the "significant event" in which the client is asked to recall some short period (30 to 60 minutes) from the recent past during which something relevant happened to her or him. Detailed recollections of actions, thoughts, and feelings are to be recorded in writing or by use of a tape recorder. As a second step, after some break, the client is to reread or play back the first account and record additional memories, comments, and reactions. The final task then is to formulate a meaning for the material generated so far. A good deal of insight into the dynamics of the client's situation can be gained this way, which may eventually lead to fresh solutions.

My preferred approach to the study of burnout can be summarized as follows. Because I reject the possibility that people are born burned out, it must be experiences later in life that cause them to burn out. So let's look at whatever these experiences are. When we meet an individual in an advanced stage of burnout, something must have happened to her or him previously. And what has happened will have had antecedents. And before that, there will have been yet another story. When to stop all of this root searching will be partly a matter of convenience and surely one of judgment. But I suggest that at the beginning of a burnout process we will find, inevitably, some frustration or loss of autonomy with which the individual failed to cope in an adequate way.

AN ACTION MODEL OF BURNOUT

Space limitations permit only the roughest sketch here of a model that I have detailed at some length elsewhere (Burisch, 1989). As is customary in action theory, my basic unit of analysis is an action episode (AE) of unspecified extension, ranging from minutes to decades. AEs can be hierarchically nested. Matters are simplified by analyzing AEs one at a time, although in reality most people typically find themselves in several AEs concurrently.

The Undisturbed Action Episode

An AE begins when one or more of an actor's latent motives are activated by some perceived situation. The result is a commitment to an incentive, in the terminology of Klinger (1975). To reach the incentive, the actor engages in some action planning, utilizing his or her "world model" to form expectations concerning the needed time and resources, the likely benefits, and the risks of negative side effects. When the necessary steps are successfully carried out and the incentive is attained without investing more than the appropriated resources, the motive becomes temporarily satiated, the episode is considered satisfactorily completed, and the world model has been consolidated. So much for an AE without complications.

Four Types of Disturbed Action Episodes

More pertinent to the study of burnout is the case where things do not work out that smoothly. Some obstacle may interfere with goal attainment, either calling for unexpectedly high investments (*goal impediment*) or blocking the goal altogether (*motive thwarting*). Alternatively, the goal may be attained, but the reward fails to live up to expectations (*insufficient reward*). Or, although everything looked fine at first, *unexpected negative side effects* may offset much or all that had been gained.

The first two of these four types of disturbed AEs will be illustrated by excerpts from what to my knowledge is the earliest and best documented case in the professional literature, Schwartz and Will's (1953) analysis of short-term burnout in a psychiatric nurse. Miss Jones, the protagonist, works on the chronically disturbed ward of a mental hospital. Despite the fact that "favorable changes in patients are slow, and when these changes take place, they are often barely perceptible," it is said that "personnel throughout the hospital attempt to provide a therapeutic milieu by maintaining enthusiasm and interest in their work despite very great difficulties." The story begins when Miss Jones returns, after a brief absence, "with enthusiasm, eager to resume her work," only to find the ward in a crisis of "low morale." The regular charge nurse has been replaced by a substitute, there is a shortage of personnel and a general feeling of fatigue, discouragement, and lowered effectiveness. When Miss Jones tries to intervene, she experiences one or more episodes that can be characterized as motive thwarting (failure): "Miss Jones initially thought that by offering suggestions for the improvement of the ward routine and patient care, she might help to bring about some alteration in the effectiveness of the ward personnel. Some of her suggestions were met with little enthusiasm; others were met with opposition and resistance, because of the low morale prevailing in the group." Following these unsuccessful attempts to improve the functioning of the ward, the nurse decides to do what she can do on her own: taking care of individual patients. But there is plenty of goal impediment in these AEs, because meanwhile the patients have begun to react to the staff's demoralization: "Miss Jones came more and more to experience the patients as being very difficult to deal with. Withdrawn patients seemed to withdraw more from her; aggressive patients became more aggressive; some patients became more demanding; other patients became panicky; and many were resistant and negativistic when Miss Jones approached them." The paper reports detailed observations which, to me, leave no doubt that a prototypical burnout process was set in motion.

Examples of the third type of disturbance, insufficient rewards, are easy to find in the burnout literature. Teachers complain that no matter how creatively they teach their classes, all that is forthcoming is rejection (Bardo, 1979). Among the factors invoked by Mohler (1983) to explain burnout in his intensive study of air traffic controllers is management's indifference to "individual recognition of especially well-done work." Aronson, Pines, and Kafry (1983) note that dentists practically never get positive feedback for their work; patients only call when something goes wrong.

Finally, negative side effects are what bring people into various types of conflict. Consider the pipefitter who is caught in the middle "between union and management. Can't do electrical work like management wants—I have to get an electrician to do that or break the contract with the union." Kahn et al. (1964, pp. 57–59) have any number of additional examples.

First- and Second-Order Stress

A single disturbed AE is rarely sufficient to cause burnout (although there must always be one first experience). Rather, it may cause stress—"first-order stress," as I prefer to call it, in order to distinguish it from "second-order stress," which results from failing attempts to remedy the situation. In Farber's (1983b, p. 14) words, "Burnout is more often the result not of stress per se . . . but of *unmediated stress*—of being stressed and having no 'out', no buffers, no support system."

Consequences of Second-Order Stress

Coping with second-order stress and the concomitant loss of autonomy may be successful and eventually may lead to growth in the sense of a more realistic world model and enhanced competence in handling the world and oneself. Coping may also fail, however, and this failure may trigger a burnout process.

Unsuccessful coping, if severe or frequent enough, can have any or all of the following consequences, plus some others. The motive profile may change, with some motives (such as being an effective helper) being first inflated beyond all bounds, then extinguished or countered by strong avoidance motives. Certain motive-activating situations that had formerly been sought out may be shunned to prevent a motivational conflict. Action planning may become excessively perfectionistic, inadequate because of panic, unduly sloppy, or replaced by quasi-automated reacting. Level of aspiration may shift downward. The ability to reach one's goals may suffer in general. The world model may be shaken profoundly, with repercussions for all previous reaction modes. People may give up early where fighting would succeed, or go on fighting long-lost battles. The feeling of self-efficacy (Bandura, 1977), the expectation not to be helpless vis-à-vis the world, may give way to what Frank (1961) called demoralization and pinpointed as the common characteristic of people who seek psychotherapy.

Finally, certain ways of coping, such as withdrawal or drug use, may be adopted in place of more promising ones, making bad situations worse. Applying the action episode model to individual burnout biographies usually reveals a number of such vicious circles or positive feedback loops. Schwartz and Will's (1953) case of Miss Jones contains many examples, the most pervasive one being a troubled personal relationship leading to withdrawal on the part of one or both partners, which in turn causes the relationship to deteriorate further.

Burnout Proneness

Let me end this discussion of the action model by pointing out that it also offers a frame of reference and useful categories for the dispositional side of burnout, i.e., for what makes people burnout-prone. For instance, the motive profiles of burnout-prone individuals may contain certain motives that are by themselves detrimental. Proving to oneself and the world that one is no longer a weak, needy child may be one such motive (Fischer, 1983; Schmidbauer, 1977). It is possible that burnout-prone individuals, at the action planning stage of superordinate AEs, systematically overestimate the happiness-generating power of successful AEs while systematically underestimating the not-so-obvious costs incurred. That would explain "top-of-the-hill blues" following, at long last, being appointed chairman of the board, being given a professorship, or winning a Prix Goncourt(Carrière, 1987). Moreover, vulnerable individuals may be inclined initially toward extremely high ambitions (Boy & Pine, 1980), rigid life scripts (Forney, Wallace-Schutzman, & Wiggers, 1982), or a world model overly dominated by the just world hypothesis (Lerner, 1980)—dispositions that considerably raise the chances of eventually encountering a string of important action episodes that fail. At a more tangible level, training deficits (e.g., lack of skills for goal attainment) have been invoked to explain burnout proneness (Heifetz & Bersani, 1983).

CONCLUSION

The model just sketched does not readily lend itself to empirical testing because a high degree of "verisimilitude" (Meehl, 1978) was the chief criterion guiding its construction. It is a conceptual framework, a vehicle for thought, rather than one of the simplistic theories stating that burnout is the consequence of some factors X, Y, and Z. Whoever hopes to do a quick thesis on it is bound to be frustrated. I must leave it to the reader to judge whether this is an asset or a liability.

In summary, I have presented here what I believe is the foundation for a comprehensive burnout theory, to eventually supersede a lot of the laundry list approaches of the past (Shirom, 1989). Whether we speak of incentives or resources, and of autonomy or control or conservation, is in large part probably a matter of linguistic preference. To repeat, my main point is that at the *molar* level burnout has been elucidated fairly well by several research traditions, basic terminological fuzziness notwithstanding. Thus we do not have to start from scratch or reinvent the wheel. At the *molecular* level we are at best getting ready to start. However, the action model outlined above should serve as a point of departure.

6

BURNING OUT: A FRAMEWORK

Lennart Hallsten
National Institute of Occupational Health, Sweden

Recent methodological improvements within the area of burnout have permitted a more solid foundation for the burnout phenomenon (cf. Garden, 1987a; Golembiewski, Munzenrider, & Stevenson, 1986; Jackson, Schwab, & Schuler, 1986; Keinan & Melamed, 1987; McCranie & Brandsma, 1988; Melamed, Kushnir, & Shirom, 1992; Schaufeli & Van Dierendonck, 1992; Shamir, 1986; Wade, Cooley, & Savicki, 1986; Wolpin, 1988). In his valuable review, Shirom (1989) delineates and discusses many conceptual consequences of these later studies. He concludes that burnout essentially "refers to a combination of physical fatigue, emotional exhaustion and cognitive weariness" and that "the depletion of energetic resources . . . does not overlap any other established behavioral science concepts" (p. 33). Burnout is a chronic, negative, affective response with fatigue and emotional exhaustion as its core aspects. This view of the burnout phenomenon is here referred to as the state conception of burnout, since it identifies burnout with affective states. Researchers within the field may have different opinions regarding antecedents and consequences of the phenomenon, but most of them appear to adhere to this state conception and to the view that burnout is adequately operationalized by the Burnout Measure (BM) (Pines & Aronson, 1988) or by the emotional exhaustion scale from the Maslach Burnout Inventory (MBI) (Maslach & Jackson, 1981a). Shirom seems to be an exception in stating that scales such as MBI and BM are contaminated by depression, frustration, and anxiety and that these scales should be purified.

Most of the conclusions from Shirom's review are reasonable. However, the review does raise some questions and doubts. Nearly all studies referred to are based on MBI and BM or variants of them, and consequently the accuracy of the conclusions is dependent on the validity of these instruments. Significant data from depth interviews or clinical settings (e.g., Firth, 1985; Fischer, 1983; Freudenberger & North, 1986) are lacking, and the quite different portraits given from these and other contexts (active and passive burnout; chronic and temporary states; acute and progressive onsets, etc.) are not commented on. Finally, the inference that the gradual depletion of resources does not overlap with other established concepts seems questionable.

The concept of burnout was earlier criticized for being mainly descriptive, anecdotal, and vaguely defined (Farber, 1983a; Freudenberger, 1983a; Maslach, 1982c; Meier, 1983). This criticism is still valid, but perhaps not of primary significance. Presumably, the basic problem is that burnout does not have a sufficiently distinctive character in comparison with such related concepts as depression, stress, and alienation. Its etiology and its distinguishing aspects in relation to these phenomena are not specified. In parallel to what Abramson, Metalsky, and Alloy (1989) noted about depression, it is argued here that definitions of burnout based on symptoms alone are less useful than those based on an etiology, not just for conceptual clarity but also for prevention and cure.

Since burnout, as measured by MBI or BM, has been shown to correlate with nearly every organizational and psychological aspect of the work environment (Shamir, 1986), researchers run the risk of blind empiricism if models and frameworks are not formulated. The requested models should include specific etiological factors or mechanisms for the phenomenon, and comparisons with more validated concepts, such as depression, should be included. In addition, our interpretations should be consistent with data from both inventories and clinical settings. Last but not least, we should try to state what burnout is not.

The state conception of burnout and its validity is here briefly examined and questioned. The conclusion reached is that the state conception of burnout makes it overinclusive. A framework for the burnout phenomenon is then put forth where the focus is directed toward a more precise phenomenon, the process of "burning out," which eventually may progress to the state of burnout. The purpose for this framework is to distinguish burning out from similar phenomena and to outline guidelines for operationalizations and prevention.

SOME WEAKNESSES OF THE STATE CONCEPTION OF BURNOUT

Studies regarding the discriminative validity of burnout, as measured by MBI and BM, have shown that it is difficult to distinguish it from depression (Firth,

McIntee, & McKeown, 1985; Meier, 1984), poor mental health, and psychological strain (Lee & Ashforth, 1990). Emotional exhaustion and depletion of resources appear to be close equivalents to "lowered energy and chronic fatigue," which are generally regarded as criteria for depression syndromes (dysthymic disorder) (American Psychiatric Association, 1980). Other diagnoses may coincide almost exactly with the state conception of burnout, e.g., emotional fatigue (Laughlin, 1967). Consequently, depression and emotional fatigue seem to overlap in the state conception of burnout.

In a Swedish human service agency the prevalence of burnout among clerks, ancillary workers, and those in human service professions was studied (Hallsten, 1988). This study was based on a small sample ($N = 60$) but had the advantage of a very low nonresponse rate (8%), which is important since nonresponse can be a sign of burnout. Presumably, burnout is generally conceived to be more prevalent among client staff than among clerks, ancillary workers, and persons in low organizational positions with highly restricted decision latitudes. These latter groups might be expected to be more exposed to "mental strain" as was shown in earlier studies of occupational health (e.g., Karasek, 1979). It was found that clerical and ancillary staff had much higher levels of BM than the client staff. More than 30% of the clerical/ancillary staff had BM values above 4.0 compared to 10% of the client staff. Although this outcome may not match the general preconceptions of burnout, it does correspond to Shirom's view that burnout is a ubiquitous experience. However, this result is also consistent with an interpretation of BM as an index of mental strain.

The state conception of burnout does not seem to coincide with connotations from everyday language. In previous studies that have characterized persons prone to burnout, at least two different types have emerged: a committed, idealistic, obsessive professional (e.g., Farber, 1983b; Freudenberger, 1974) on the one hand and a "weak, unassertive [person] . . . anxious, fearful of involvement, . . . lacks self-confidence, has little ambition" (Maslach, 1982a) on the other. In my opinion, only the former description coincides with the intuitive notion of burnout. At least in Swedish, the term burnout (*utbränd*) connotes a state preceded by internally generated, intense strivings toward certain goals. Consequently, it is suggested that the term burnout should refer to the case where exhaustion is preceded by commitment and pseudoactive strivings, while perhaps the alternative case might be labeled depression or being worn out (Fischer, 1983; see also Chapter 11). Use of the terms *active* and *passive* burnout (Golembiewski & Munzenrider, 1988) for these categories of reactions has the drawback of monopolizing the field and leaving a well-established concept such as depression with no denotation.

The state conception of burnout may conceal the relationship between burnout and environmental factors. Although personal characteristics are generally assumed to contribute to burnout, to my knowledge only Garden (1987) presented data concerning the moderating influence from these aspects on the relationship between environmental factors and burnout (on the MBI dimension depersonalization). In the interview study referred to earlier (Hallsten, 1988), subjects were classified according to ego developmental models (Kegan, 1982; Loevinger, 1976) into autonomous and socially oriented employees. Separate analyses were made for these two groups for the regression of BM scores on various aspects of perceived job aspects and attitudes.

It was found that these two groups differed in many of the relations between BM scores and perceived environmental aspects. Burnout was not related to the same environmental factors or to preferred interventions for these two groups. Combining the data from these two groups concealed these relationships and often rendered them statistically nonsignificant. For instance, among autonomous persons, the higher their BM scores the more they welcomed organizational interventions to alleviate burnout, while the opposite was true among the socially oriented group. Because this division into socially and autonomously oriented groups was clearly correlated with commitment and coping style, a more narrow definition of burnout may be less influenced by these personal orientations.

To summarize, the state conception of burnout, with emotional exhaustion, fatigue, and depletion as its central criteria, appears inadequate for the differentiation of burnout from similar concepts such as depression. It seems to be overinclusive in that it clusters different behaviors and attitudes under the same label, disregards etiological aspects, may inflate or conceal relations with environmental factors, and is not in agreement with everyday language. Emotional exhaustion has perhaps the same status as fever and headache as being a definite symptom of a disease but insufficient as a distinguishing criterion for a certain phenomenon. A phenomenon requires an etiology, certain symptoms, and a development.

Some potential advantages in using the concept of burnout are that it may capture many of the problems within human service professions and that it may add something new to more validated concepts such as stress, strain, and depression. These advantages should be retained.

AN OUTLINE OF THE PROCESS OF BURNING OUT

The following framework has some advantages as compared to state conceptions of burnout. First, it is in agreement with the intuitive notion of burnout as delineated above. Second, it elaborates the process and coping pattern resulting in burnout as well as possible exits from this process. Third, it can account for crucial outcomes from studies of burnout, such as the relative stability of burnout and its relatedness to earlier assessed personal characteristics (McCranie & Brandsma, 1988; Shirom, 1986; Wade, Cooley, & Savicki 1986) as well as the differential impact of situations on depressive reactions (Barnett & Gotlib, 1988; Brown & Harris, 1978). Fourth, it facilitates the discriminative validity of the concept. Fifth, it offers a foundation for differentiated countermeasures to burnout and similar states. In this framework, the focus is shifted from the burnout phenomenon to the process of burning out. The occupational sector for professionals constitutes the context for the process as described here, but it might apply in any context where deep and long-term commitments appear. Because the coping pattern is considered essential to burning out, the framework outlines the process as an individual phenomenon; however, this does not imply that the contributions to the process have mainly an individual origin.

The standpoint taken here is that burnout is a certain kind of depression. Depression is assumed to occur when the enactment of a self-definitional role is disrupted and when no alternative role is available. A self-definitional role is "primary in providing the basis for a person's sense of self" (Oatley & Bolton, 1985). Both depression and burnout imply negative views toward the environment, one's own person, and the future (Beck, 1976), as well as exhaustion and depressed mood. As such, the state of burnout is not a unique phenomenon although it might show some minor dissimilarities to various expressions of depression. The interesting aspect of burnout is its etiology. Burnout is a form of depression that results from the process of burning out, which is a necessary cause of burnout. Hence, burning out is one route to depression. *Burning out is assumed to appear when the enactment of an active, self-definitional role is threatened or disrupted* with no alternative role at hand. This is in accordance with the intuitive notion of burnout.

The structure and phases of the burning out process are presented in Figure 6-1. The phases of "absorbing commitment" and "frustrated strivings" constitute the core aspects of the process that may, but need not, evolve into some form of depression or solution. Frustrated strivings exhibits the most distinctive aspects of the burning-out process and forms the basis for a short-hand charac-

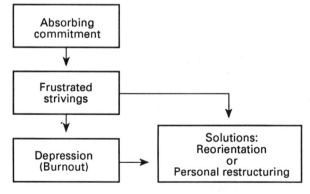

FIGURE   6-1 The process of burning out.

terization of the process. Burning out refers to the recurrent, interactive pattern of anxious-depressive reactions and rigid, pseudoactive strivings. This pattern, which eventually may result in exhaustion, fatigue, and depressive episodes, is the unique part of the burnout phenomenon to be explained and understood. These strivings have obsessive, boundless features paired with denial, and self-preservation may eventually be suppressed. The burning-out process has consequences that bear resemblance to the crisis in that it may result in both negative and positive outcomes.

Burning out may occur in any occupational or nonoccupational context and a hypothetical example is here given from a research setting. A scientist invests much of his professional interest and self-esteem in a series of studies. Eventually, some severe flaws in the design are discovered and the scientist becomes preoccupied with making the best of the situation (absorbing commitment). He cannot hope for new grants for new, redesigned studies and he cannot conceive of support from his colleagues. He finds himself forced to complete the studies by utilizing his available resources and talents. Consequently, he devotes all of his energy to laborious and intricate analyses in the hope of saving his studies and his self-esteem. Although he repeatedly fails, with subsequent increases in frustration and pessimism, he cannot abandon his idea and strivings to improve the situation on his own (frustrated strivings). This coping pattern is very exhausting, and it may eventually compel him to give up his efforts, resulting in an enduring depressive episode which may be called burnout.

KEY CONTRIBUTING FACTORS TO BURNING OUT

Three interacting, hypothetical constructs are introduced to account for the burning-out process: a vulnerability, a goal orientation, and an incongruous, threatening environment.

Vulnerability

A certain degree of personal vulnerability seems to be a necessary prerequisite for burning out. As indicated by the labels "absorbing commitment" and "frustrated strivings," the person is significantly preoccupied with a present situation that is regarded as vital. Few options to create a positive sense of self are perceived, and consequently the person is highly dependent on an available self-definitional role and on its performance. Presumably, the person has had some troubling experiences (such as low social support and few possibilities for satisfactory relations) that make his or her self-esteem sensitive to recent experiences. This vulnerability moderates the interpretations of, and reactions to, problematic situations. For instance, alienating conditions are more likely to result in depressive episodes for vulnerable persons than for low-vulnerability persons. Vulnerability may also account for the somewhat invariable character of burnout as noted in longitudinal studies.

The degree of vulnerability might be defined by the following related indices: (1) the degree of instability of self-image and self-esteem, (2) the degree of dependence on self-definitional role enactment and the lack of subsidiary or potential roles for self-definition, and (3) the degree of social support outside the present work domain. Lack of social support is probably the basic factor contributing to vulnerability. Vulnerability is presumably correlated with self-complexity and multiple identities (Linville, 1985; Thoits, 1983b). Lower self-complexity is related to higher vulnerability and greater swings in mood (Linville, 1985). Other concepts assumed to be indicative of this dimension of vulnerability include negative affectivity (Watson & Pennebaker, 1989), negative cognitions and attributions (Abramson, Seligman, & Teasdale, 1978; Beck, 1976), sense of coherence (Antonovsky, 1987), and hardiness (Kobasa, Maddi, & Zola, 1983).

Goal Orientation

Another essential aspect of this process is the active orientation toward long-term goals (e.g., improve organizational routines, help certain clients, etc). This is coupled with a reliance on one's own actions and responsibility. Hence, the person takes the initiative and tries to solve problems. He or she expects to be active and is driven by ambitions. However, this alleged commitment often conceals various defensive aims that can make the actions narrow and misdirected. This is obvious in frustrated strivings, which might appear as maladaptive. The interpretation given here is that these strivings constitute a part of an active self-definitional role enactment under threat; that is, they have the function of creating or maintaining an acceptable personal identity and meaning in life. The person does not withdraw from the situation because this would produce a more devalued outcome than continued strivings. Self-preservation is not pursued because the person feels that at present he or she has no respectable self to preserve. This active, or pseudoactive, pattern may continue as long as the person has not totally given up the possibility of creating positive self-esteem through role enactment. In the latter case, the vulnerable person may act passively, perhaps with a pessimistic, detached, wait-and-see attitude and with negative views of both self and surroundings.

The degree of goal orientation can be estimated from the degree of (1) commitment expressed and (2) effort displayed regarding long-term goals. Occupational responsibility and professional training are assumed to be positively related to this goal orientation. Vulnerability and identity strivings may influence this goal orientation by making it more intense, rigid, and misdirected.

Perceived Environmental Congruency

The third key aspect for this process is an environment that is perceived as incongruous. For professionals, the availability of two organizational factors is crucial: (1) personal and organizational competencies and resources for attaining organizational goals and professional standards, and (2) various forms of social support, including shared values and goals. Both these factors seem necessary to the vulnerable professional to create a positive self-image. The more incongruous the organizational environment, the lower the expectations of achievement and high self-esteem. Without organizational problems, the process of burning out is unlikely to occur for professionals.

The degree of perceived environmental congruency corresponds to (1) perceived personal and organizational competencies/resources for attaining organizational goals and professional standards, and (2) perceived social support and shared goals. Goal expectations are primarily related to this congruency. Vulnerability and low self-complexity may contribute to more modest and unstable, reactive perceptions of congruency (Linville, 1985). Perceived low congruency usually accentuates the vulnerability.

BURNING OUT AND RELATED PROCESSES

This burning-out process is consistent with some control, cognitive, and behavioral models of reactive depression (e.g., Oatley & Bolton, 1985; Pyszcynsky & Greenberg, 1987). As in these models, the occurrence of burning out is presumed to be highly dependent on a predisposing factor of vulnerability. This vulnerability may be displayed in the investment of self-worth into a single role, and it is the threat to or the loss of this role that is assumed to start the burning-out as well as the depressive process. In contrast to these two models of depression, this framework of burning out emphasizes the achievement motives rather than the affiliative motives involved. The burning-out process can also be conceived of as an instance of narcissistic activity (Kohut, 1971; Morrison, 1986; Stolorow, 1986), as already noted by Fischer (1983). Models of narcissism may offer a basis for understanding burning out, and Cooper (1986) suggested that narcissistically disturbed persons have a tendency to burn out.

This framework for the burning-out process was derived ad hoc and has not been directly tested in any study. It is an attempt to integrate data and models from stress, depression, and burnout research areas as well as from different methodological approaches. Important contributions also stem from empirical data (Hallsten, 1985, 1988) and from my informal discussions and observations of staff behavior in human service organizations over many years. The framework offers a crude structure with the primary purpose of contrasting burning out to related processes. This means that the contrasting processes are treated superficially without sharp distinctions.

Individual change processes are conceptualized here in terms of "phases" (for lack of a better term), although it is well known that phase models may be difficult to validate (see also Chapter 14). There has only been equivocal support for phase transitions as responses to undesirable events (Silver & Wortman, 1980). Consequently, the number of phases here has been reduced to a minimum. The framework has similarities to and has been influenced by other models (Bowlby, 1979; Edelwich & Brodsky, 1980; Klinger, 1977; Viney, 1976; Wortman & Brehm, 1975). However, in contrast to other phase models, no necessary progression from one phase to the other is assumed (cf. Shirom, 1989), as indicated by the broken arrows in Figure 6-2. An overview of the framework is presented in Figure 6-2, where four types of development are related.

As discussed in the previous section, the framework is based on the three interwoven dimensions of vulnerability, goal orientation, and perceived environmental congruency. These dimensions are used to differentiate phases, which are the dominant cognitions, attitudes, and coping patterns that people may exhibit in response to work, self, environment, and future. It should be noted that the four downward careers to be sketched stand in contrast to an ideal process that might be called "gradual growth" (not illustrated in Figure 6-2). This upward career would comprise adequate tasks, resources, and organizational goals; perceived mutual support and trust; genuine communication; and so forth. The four downward careers describe what happens when this positive fit between person and job does not occur. The assumed phases and their characteristics are briefly presented below.

Absorbing Commitment

The predisposing factors that set the stage for this phase are (1) a certain vulnerability in the person, (2) an ambiguous and somewhat problematic job that (3) has generated some expectations of an active self-definitional role enactment. After an initial outlook of wait-and-see, the vulnerable person may exhibit an attitude of absorbing commitment that is assumed to be a necessary but insufficient precondition for the process of burning out. This attitude emerges when the promising job is seen to entail some troubles that increase preoccupation and ambivalence. Expectations, self-esteem, and self-image seem highly reactive and variable. The self-conception is loosely integrated, which sometimes makes coping inadequate to true needs and capacity. This attitude of absorbing commitment has certain similarities to the phase of enthusiasm, which Edelwich and Brodsky (1980) assumed to be the outset of the burnout process. The assumed anxiety and ambivalence coupled with the commitment are, however, more pronounced here.

The characteristic signs of this phase include the following:

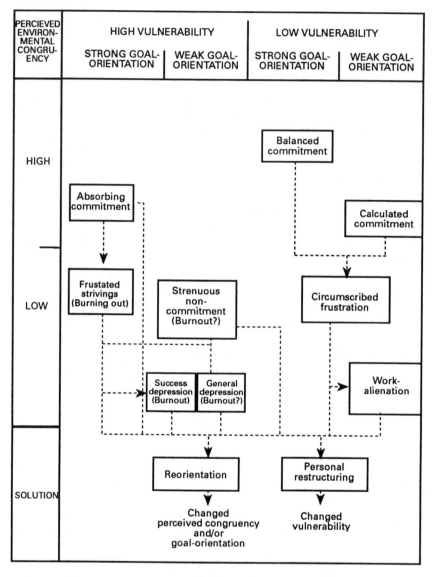

FIGURE   6-2 A framework for burning out compared to similar processes.

1. Perceived options to create meaning in life and to confirm a positive personal identity by accomplishments in, or by outcomes from, the job; an alleged commitment to organizational goals (for professionals);
2. Indications of concern, responsibility, initiative, energy, and willingness to identify with work, task, and organization; search for a mode or a position

to offer a significant or unique contribution to the organization (or to the invested object); unsure of support and own capacity, mild pseudoactive strivings;

3. Certain anxiety and ambivalence toward the job; fluctuating and reactive expectations, self-esteem, and self-image; preoccupation with work problems and personal future;
4. Activated scripts and coping patterns.

Indications of influences from scripts, automatic thoughts (Beck, 1976), or schemata (Segal, 1988) come from an extensive study on burnout in a Swedish human service organization (Hallsten, 1985, 1986). As part of this study, 17 professionals were selected (on nonformal grounds unrelated to this framework) for exploratory, depth interviews about their reactions to work-related problems. These persons had earlier been candid about their frustrations in the organization, and they could recognize episodes similar to burning out in their earlier or present jobs. Some recurring themes emerged in these interviews: the view of the present job as a possibility to avenge oneself for earlier critical experiences or failures; strong identification with the clients and a sense of obligation to help them; the abundant obstacles and inadequacy of the organization; preoccupation with improvement of work procedures. These themes appeared to reveal some semiconscious expectations and values acting as motivators or scripts for these professionals.

The most crucial script seemed to concern personal identity and may be formulated as:

1. "I am my achievements." In other words, identity and self-image are contingent on the outcome of future achievements. This identity script, which summarizes the need for an active self-definitional role enactment, is assumed to represent the key motivational factor in the burnout process. In addition, some other script-like expectations and values displayed by these professionals in various degrees may be expressed as:
2. "I should follow the rules and behave fairly in my job" (the ethical aspect);
3. "This task/job is important to others as well as to me" (the work value aspect);
4. "I should be able to accomplish substantial results here although I foresee certain problems" (the competence aspect);
5. "I should be given (I am entitled to) control, support and rewards in proportion to my commitment" (the reward aspect);
6. "Hence, I should be able to lead a rich and good life" (the future aspect).

These scripts are assumed to summarize the motive structure of absorbing commitment for professionals. Nonprofessionals prone to burnout expressed more instrumental attitudes. Work values and occupational accomplishments per se were often of less importance for this latter group than, say, the wish to create a good life for their family through their job earnings. The five last scripts/assumptions appear somewhat idealistic but they seem to be embraced by many other employees in less anxious forms (cf. Pines & Aronson, 1988). Eventually, these scripts may be modified by various work experiences, often of a negative nature, that may accelerate the burning-out process. From a phenomenological perspective, the process of burning out may be conceived of as the questioning and loss of these scripts with, perhaps, a turn to opposite views (with the exception of the identity aspect). First the expectations are questioned ("I am not able to accomplish substantial results"), and later the values ("This job is unimportant to others as well as to me"). Hence, burning out means modifications in perceived competence, reward, future outlook, moral attitudes, and work values, while a negative self-image seems gradually confirmed.

The attitude of absorbing commitment may unfold in different directions. One is the negative route described above, with strenuous encounters in work subsequently leading to frustrated strivings and burning out. Another option is job turnover (reorientation), and a third alternative is a gradual, positive growth, mainly resulting from a congruous environment (gradual growth). Finally, a fourth but unlikely possibility is a prolonged stay within this phase. In the latter three cases the person should not be regarded as burning out or burned out. The process need not proceed from this or any other phase.

Frustrated Strivings (Burning Out)

If the expectations presented above are not realized and the mild pseudoactive coping style fails, the person may continue to the phase of frustrated strivings, where the label of burning out is adequate. Critical professional-occupational failures, obvious lack of social/occupational support, discarded job offers, psychosomatic symptoms, and the like may have the effect of bringing the person to this phase. The salient aspect here is that the person feels obstructed from attaining his or her goals. In spite of this, the strivings and coping efforts are recurringly pursued because alternatives are lacking and inactivity evokes fear. An interactive pattern of activity and anxiety-depression is noted. This is the distinctive phase of the burning-out process.

It is in this phase that we see the closest similarities to the narcissistic pattern as described by Fischer (1983), who might describe these strivings as the pursuit of the illusion of grandiosity. In parallel to what Pyszcynsky and Greenberg (1987) outlined for reactive depression, this phase can be seen as an attempt to keep or regain a lost object that functions as a source of identity and self-esteem. A difference is that this attempt is presumed to result in pseudoactive strivings, not just in self-focusing, rumination, and withdrawal. Klinger (1977) argues that the typical reaction to frustrations is invigoration, which is similar to the frustrated strivings described here. He suggests that the loss of perspective can explain the increased effort displayed to reach the goals.

The characteristic signs of this phase include the following:

1. Cognitions of lost control and powerlessness; perceived threats to the link between means and ends; increased ambivalence toward work, staff, and clients; increased fears of appearing weak and worthless; no available coping alternatives; awareness of the harmfulness of own coping style; occasional improvements in situation and outlook; lowered self-esteem but the potential, positive self-image is not totally abandoned; selective cynicism; more pessimism about the future
2. Competence, reward, and future expectations nearly lost; ethics and work values are questioned while the identity script remains intact and reinforces conflict
3. Recycling efforts; occasional withdrawals; efforts to conceal the detrimental mental state; strivings appear to arise both from occasional improvements in the situation and from threats such as being regarded as weak, irresponsible, worthless; harmful effects of the coping pattern per se
4. More permanent emotional exhaustion, disappointment, anger, loss of energy, depressed mood; intensified rumination

This phase is perhaps the most distressing part of the burning-out process, but it seems that many can live with this heroic attitude for years. A considerable percentage of professionals eventually find themselves in this phase partly in response to obvious inconsistencies in their work organizations. Later on, this attitude of frustrated strivings may (but need not) transform into other attitudes such as success depression, general depression, reorientation, or personal restructuring.

Success Depression (Burnout)

In frustrated strivings the person has failed to reach his or her goals due to external obstacles and/or a lack of personal resources. The outcome expectations were mistaken. However, a phase where the label of burnout is adequate is "success depression," which results when the person actually reaches expected outcomes but the expected rewards do not materialize. In this case, it is the reward expectations that were mistaken. This state is similar to completion depression described in other contexts (e.g., Laughlin, 1967). The characteristic signs of this phase include the following:

1. Characteristics similar to those of general unipolar depression, such as depressive mood, low energy, hopelessness, pessimism (e.g., Abramson et al., 1978, 1989; American Psychiatric Association, 1980; Beck, 1976.)

2. Lost expectations; low self-esteem; ethics and work values deeply questioned

Because the goals have been reached, anger is less outer directed than in frustrated strivings. The label of burnout may be adequate here because self-esteem and meaning in life seem to be lost after hard strivings. However, because the social status is high and the person is compelled to admit his or her own erroneous expectations, this may facilitate a later solution. The successful but burnt-out architect in Greene's (1960) novel may be a representative of this phase. Because of the organizational intricacies, this phase is less common than frustrated strivings for personnel within the helping professions. This attitude of success depression may evolve into attitudes such as reorientation or personal restructuring.

Strenuous Noncommitment (Burnout or Entrance to Depression?)

The phase of strenuous noncommitment usually arises when the vulnerable person feels constrained to adapt to a new, unattractive job with no other options at hand. Neither this job nor any other is expected to function as an adequate arena for identity creation. This phase may also be a "solution" to a preceding burning-out process whereby the person tries to detach from earlier self-definitional strivings and attitudes (reorientation). This induces a passive coping style, and most expectations and values of absorbing commitment are lacking, although the person is highly dependent on external sources for self-esteem.

The characteristic signs of this phase include the following:

1. Positive expectations and values are deferred; job offers no possibility to create an acceptable personal identity; self-evaluative involvement in the job and identification with work, staff, products are low; detached attitude; perceptions of control in the work but not over it; expectations to conform to acceptable performance standards and to get fair treatment from the organization
2. Moderate efforts to get the job; passive coping to handle problems in work; blame and tacit criticism are common; inclination toward cynicism-nihilism
3. Feels more or less compelled to take the job; vague dreams and distressed preoccupations about job future without plans to reach them; detachment causes dissatisfaction and distress because the lack of closeness and support cannot be compensated for; more irritation and strain than depressed mood; low personal pride; emotionally but not socially alienated; not a total victim

Usually, this outlook is the debut of a depression or wearing-out process, and it may be regarded as an instance of masked depression because strain, irritation, and detachment tend to conceal depressive feelings. Klinger (1977) also describes depression as a possible consequence of prolonged invigoration. Persons within this category may get high scores on burnout inventories, but the label burnout should not be applied unless this phase was preceded by absorbing commitment or frustrated strivings. Many traditional burnout portraits seem to belong to this phase.

This attitude toward a job can persist for a lifetime and form part of a personal character. Under certain circumstances it can lead to positive solutions through reorientation or personal restructuring. However, it can also evolve into general depression due to conditions that further tax the resources of the person.

General Depression (Depression or Burnout?)

Usually, the reason for transition to this phase is a serious blow to self-esteem and to meaning in life. For instance, a situation may occur in the organization that indicates that the person is not valued and is regarded as incompetent or as a nobody. Other common reasons for entrance into this phase are signs of illness or some severe conflict with a significant person. Pyszcynsky and Greenberg (1987) argued that a serious loss for a person may force him or her into a passive and rigid self-regulatory cycle that may result in a reactive depression.

The characteristic signs of this phase include the following:

1. Similar, but accentuated, signs of success depression
2. Expectations lost; ethics and work values deeply questioned or turned to their opposites; very low self-esteem

The person feels victimized, emotionally and socially alienated, and has no long-term goals toward which to strive. The need to attend to work diminishes, often resulting in extended sick leaves or positional transfers. Of course, people in this phase usually get high scores on burnout inventories if tendencies to social desirability (Crowne & Marlowe, 1964) do not interfere. The designation of burnout may be accurate here, but if no traces of the preceding attitudes of absorbing commitment and frustrated strivings are noted, then persons should be classified as worn out or depressed.

Positive Changes: Reorientation and Personal Restructuring

Both the burning-out and the depression processes may eventually result in more sudden, positive changes. For example, the same object/job may be modified (e.g., through job reorganization), and this may be a beginning of gradual growth. Other changes may result from certain coping acts that are here called reorientation. Reorientation refers to a more or less deliberate change in the attachment to activities, objects, or roles, based on their perceived fit or congruency to the person. Sometimes reorientation results in solutions characterized by weaker attachment and goal orientation, as in the development from frustrated strivings to strenuous noncommitment, but it may also evolve into more positive outcomes with stronger attachment and goal orientation. For instance, the person may turn his or her interest and self-definitional strivings to another role or activity within or outside the occupational sector. In this case, reorientation comes close to regaining a lost role that Oatley and Bolton (1985) present as a resolution of a depressive episode. However, the possibility remains that reorientation just generates a temporary recovery, depending on the actual fit between the person and the environment. Perceived environmental congruence and/or goal orientation, but not vulnerability, are changed.

If the new role is suitable, perceived threat is reduced. This may facilitate a development from absorbing commitment to an attitude of "balanced commitment." However, a necessary step is that this reorientation is eventually followed by personal restructuring (see below). If the role is a suitable nonjob role (such as leisure activities), the attitude of "calculated commitment" toward the job is a likely result. If the new role is as bad as the former one, it probably means that the recovery will be temporary and that the person will soon return to his or her previous attitudes of absorbing commitment or strenuous noncommitment.

A more stable solution is personal restructuring, which may decrease the impact of the former vulnerability. Personal restructuring implies a change in the identity assumption and in the need to act in a certain self-definitional role. Usually this solution arises after alternative actions by the person, i.e., after a definite change in coping style. For example, personal restructuring may occur after the expression of covert, genuine feelings to a significant person. Contrary to earlier beliefs, feared catastrophes or retaliations do not take place, and the person then realizes that he or she is not solely dependent on achievements or the impressions made on others. Role enactments do not totally define the individual, and thus his or her self-complexity increases. A lost role is not regained, as in reorientation, but instead the need to act in a narrow, self-definitional role has diminished. The consequences of personal restructuring and gradual growth seem to be the same, although the processes involved are dissimilar.

Development for Low-Vulnerability Persons

In contrast to these phases for vulnerable persons, a brief, hypothetical outline is presented for low-vulnerability persons and their encounters with organizational problems. Antonovsky (1987) would probably regard these persons as having a coherent view of the world.

Two different categories of low-vulnerability persons are considered. Persons with a strong goal orientation in the job are assumed to have an attitude called balanced commitment. Their assumptions and values deviate from those of absorbing commitment in the lack of the identity assumption and of the idealized, anxiety-driven qualities in the scripts. Persons showing balanced commitment have a more sincere interest in their jobs without as much self-definitional striving. Perhaps they might be regarded as enthusiasts or representatives of detached concern (Maslach, 1982a). The other category of low-vulnerability persons are those with a chiefly instrumental motivation toward the job. They show calculative commitment based on a rather explicit exchange contract between job involvement and benefits (Handy, 1985). Their job commitment is generally lower but it is somewhat adapted to the attractiveness of their present work conditions. In response to work problems, their coping style is rather passive, leaving much of the initiative to other individuals, which may be an appropriate strategy.

If severe problems are encountered, low-vulnerability persons may feel dissatisfied, depressed, angry, etc., but the frustration is usually limited in duration or intensity ("circumscribed frustration"). Their identity and sense of meaning in life are not totally threatened, except perhaps, for a short period ("transient burnout," Freudenberger & North, 1986). Presumably, they have a varied repertoire of coping acts and, in addition, a high trust in other options if a coping strategy fails. A good solution to a crisis may be reorientation because radical change in the identity assumption is not needed. Presumably, reorientation is more easily attained for low-vulnerability than for high-vulnerability persons. If occupational conditions are very bad with no possibility for adaptation, these low-vulnerability persons may experience "work alienation," with feelings of powerlessness, normlessness, and estrangement from other people (cf. Lang, 1985; Seeman, 1967; Shepard, 1972). These feelings do not usually develop into depression or feelings of severe personal failure because these persons do not regard their work as a self-definitional role. Burning out and depression may occur for low-vulnerability persons only in exceptional cases, where a very serious event or successive events disrupt more than one important role.

COMPARISONS WITH RELATED CONCEPTS

The concept of burning out as presented here deviates from related concepts such as stress, alienation, crisis, and depression in that it specifies a certain etiology with motives and a behavior pattern. The outcome of a burning-out process may be similar to states and criteria of stress-strain, depression, and crisis, although it is always a result of an enduring process. As such, burning out can be considered a special instance of these phenomena.

An important aspect is the possible nonhomeostatic solutions from the process, thereby deviating from homeostatic concepts such as stress (e.g., Appley & Trumball, 1986; Cooper & Payne, 1988; Schuler, 1980; but also Hobfoll, 1988). Thus, it resembles the crisis concept (Caplan, 1964; Reiter & Strotzka, 1977; Viney, 1976) in presuming different developmental consequences. By specifying a certain etiology and behavior pattern, the framework shows similarities to the commitment-disengagement cycle (Klinger, 1977) and to the type A behavior pattern (Burke, 1984; Ganster, 1987; Pittner & Houstone, 1980). In comparison to the latter concept, however, burning out has less the character of a trait or a typology, although it bears some resemblance to the interpretation of type A behavior as presented by Matthews (1982). She emphasizes the importance of uncontrollability, self-involvement, and ambiguous standards for the occurrence of type A behavior. From this perspective, burning out might be seen as a model for the development of type A behavior into helplessness and hopelessness.

The MBI dimensions of burnout (emotional exhaustion, reduced personal accomplishment, and depersonalization) are all easily derived from the scripts assumed to be involved in burning out. Burning out, however, is a more narrow phenomenon than burnout as operationalized by MBI and BM. It is assumed that the state conception of burnout as measured by MBI and BM might include all persons belonging to the phases of frustrated strivings, success depression, strenuous noncommitment, general depression, circumscribed frustration, and work alienation, irrespective of their etiological pattern.

The only empirical data at hand, that illustrate the relationship between the state conception of burnout and burning out are from a minor study (Hallsten, 1988). There it was assumed that a BM score above 4.0 was indicative of belonging to any of the phases mentioned above. Sixteen percent of the sample had a score above this criterion (but none of these seemed to belong to success depression, circumscribed frustration, or work alienation as judged by the information presented during the interviews and the criteria outlined above). Five percent of these subjects showed a history of absorbing commitment and frustrated strivings while 11% had a career starting with strenuous noncommitment. The depression process was more common than the burning-out process. The distinguishing aspects between these two careers were primarily the goal orientation, commitment, and effort exhibited. An advantage from a perspective of discriminative validity was that the distinction between burning out and being depressed was clearly correlated with degree of professionalism. The depression process was more common among nonprofessionals while the burning-out process was more typical for professionals.

CONCLUSION

The individualistic perspective established in the above framework does not imply that burning out is regarded as exclusively influenced by individual factors. On the contrary, it is posited that this phenomenon mirrors societal and organizational aspects as well. Burning out is assumed to be an old phenomenon that has become more discernible because of societal changes. In the past, burning out was probably more likely to be found within private life. Technical developments, democratization, prolonged education, and the growing demands for openness and rationality have presumably created a foundation for burning out within such public settings as work. Discretion in occupational roles has increased, and the relative proportion of work procedures monitored by simple routines has decreased in favor of those controlled by objectives. The gradual increase in individualism within Western societies (Corbin, 1990; Taylor, 1989; Veyne, 1987) has contributed to the view that identity is no longer inherited, but is merited by one's own acts and accomplishments. All these tendencies may augment both uncertainty and achievement orientations within work. In terms of the framework presented, these societal changes should be manifested in a growing number of persons adhering to an explicit goal orientation, which implies an increasing risk of burning out at the expense of ordinary depression.

Organizational problems contributing to burning out may primarily be seen in the dimension of congruency. Presumably, the most fundamental factor contributing to burning out in our modern organizations is the gap between organizational means and ends, most notably seen in our human service organizations. Resources to meaningful ends are often missing, which can have well-known, distressing effects (goal displacement, role ambiguity, inconsistent feedback, etc.). This gap is especially taxing for vulnerable professionals.

With regard to measurement procedures, it is not enough to establish a certain state or level of burnout according to traditional criteria. The central aspects of the burning-out phenomenon lie in the need for self-definitional role enactment, in the instability of self-esteem and self-image, in the pseudoactive goal orientation, and in the perceived incongruous environment. It should be possible to develop scales for these aspects based on questionnaires and to measure them in addition to the areas covered by MBI and BM. For instance, Rosenberg's scale (1979) for perceived instability of self might be used or elaborated. Such a methodological development is now under way. The assumed presence and development of activated scripts during burning out might conveniently be explored within interviews and used for validation. Most importantly, the present framework accentuates the need for longitudinal studies of career and self-development and their influence on each other.

7

CONSERVATION OF RESOURCES: A GENERAL STRESS THEORY APPLIED TO BURNOUT

Stevan E. Hobfoll
Kent State University

John Freedy
Medical University of South Carolina

There has been a great deal of study of occupational burnout and some attempts to develop an explanatory theory (e.g., see Chapters 2–4). Despite some specific theories of burnout, however, the concept has typically been studied in a somewhat atheoretical fashion that is marked by empirical attempts to explain worker distress in a given context (Shirom, 1989).

This chapter will outline a general theory of stress termed conservation of resources (COR) (Hobfoll, 1988, 1989). This theory, it will be argued, provides an overarching framework to understand the nature of stress as a human phenomenon that is tied to people's experience regardless of the setting or context, be it work, home, hunt, or vacation. COR theory is a basic motivational theory and it is postulated that when this basic motivation is threatened or denied,

We thank Sarah Chisholm and Paula Britton for their helpful comments on earlier versions of the chapter.

stress ensues. COR theory is essentially a theory of resource utilization, and if Shirom (1989) is correct in concluding that resource depletion is a central facet of burnout, then COR theory may have particular relevance for the study of how stress leads to burnout.

In outlining COR theory, we will apply it to studies of stress in general. We will also apply its principles to research on burnout wherever possible. We will further borrow from the literature on the types of organizational stress that have been found to be related to burnout where specific studies of burnout are not available. This approach seems justified by Cobb and Rose (1973) and French and Caplan (1970). They point out that individuals in jobs where there is responsibility for others are more likely to have higher blood pressure and higher serum cholesterol, and are more often victims of stress-related diseases, such as peptic ulcer and hypertension. Thus, it would appear that burnout is a unique disorder but that it is also related to a family of disorders.

MODELS OF BURNOUT

Before discussing COR theory, we wish to briefly discuss the concept of burnout itself. Freudenberger (1974) first used the term burnout to describe a syndrome consisting of a combination of long-lasting emotional exhaustion, physical fatigue, absence of job involvement, dehumanization of the recipients of one's service, and lowered job accomplishment. This syndrome was observed among employees in chronically taxing service delivery organizations. Later, however, the concept of burnout expanded to cover all occupations where people need to interact extensively with other people on the job. Maslach (1982a) provided a more specific definition of burnout that aids its study across different professions. She defined burnout as a syndrome consisting of emotional exhaustion, depersonalization, and reduced personal accomplishment. Pines and her colleagues offered an expanded view of the syndrome (Pines, Aronson, & Kafry, 1981). They envisioned burnout as a response to stress in which there is not only a low sense of accomplishment and fatigue, but also a sense of helplessness, hopelessness, and entrapment. Cherniss (1980a) also saw burnout as a response to stress. His developmental model of burnout suggests that burnout is the last stage of a failed coping process. This final stage is seen as a defensive posture meant to halt more precipitous effects of stress.

These models share the view that burnout is a consequence of stressful work conditions. As opposed to the extreme demands of major stressors, such as the death of a loved one, burnout is a slower process. It occurs when demands are made over time in a way that tax individuals without proper rewards or resources for addressing demands. The elements of exhaustion and reduced efficacy are common to the response to more severe stress (Goldberger & Breznitz, 1982). The interpersonal aspects, however, are particular in some ways to situations where stress occurs in the context of interpersonal settings.

These may include other kinds of stress experiences, even as diverse as response to combat (Solomon, Mikulincer, & Hobfoll, 1987) or rape (Katz & Mazur, 1979). What emerges as particular to burnout is that the process occurs under a slow boil, taking time to develop. Individuals may also experience many stages of burnout and this may tend to mask the phenomenon, much as a child's growth goes unnoticed by those who see him or her daily (Golembiewski & Munzenrider, 1984).

CONSERVATION OF RESOURCES THEORY

COR theory may be applied as a theoretical model that explains the etiology of burnout and the processes that are likely to accompany chronic, relatively low-level, work-related stress. The specific motivation that is basic to COR theory is that individuals strive to obtain and maintain that which they value—these things being termed "resources." When circumstances at work or otherwise threaten people's obtaining or maintaining resources, stress ensues. Thus psychological stress occurs during one of three conditions: (1) when resources are threatened, (2) when resources are lost, and (3) when individuals invest resources and do not reap the anticipated level of return. Resources are defined as those objects (e.g., clothing, crystal goblets), conditions (e.g., employment, quality marriage), personal characteristics (carpentry skills, hardiness), and energies (e.g., stamina, knowledge, money) that are valued or that serve as a means of obtaining resources that are valued.

When considering the concept of burnout, it must also be stated that physical exhaustion or work overload (Shirom, 1989) is likely to make people feel insecure about their abilities to support the success of this motivational process. Therefore, it is further postulated that the greater salience of loss postulated by COR theory is accentuated during periods of physical or psychological overload. This occurs because (1) when many decisions are demanded, physiological overarousal is likely to ensue at the same time that (2) there are mounting doubts about one's ability to make correct judgments quickly. These latter doubts are realistic because overload decreases time available to consider a full range of options and mobilize resources, and because the complexity of interwoven problems may be beyond intellectual and organizational resources. This situation is correctly viewed as threatening and is fertile ground for losses to occur. To understand COR theory, some of the principles that follow from the basic motivational tenet need to be articulated.

Primacy of Loss

Because individuals strive to protect themselves from resource loss, loss is more salient than gain. It is interesting to speculate that this is related to survival needs that are basic to most advanced biological systems. Extra food, shelter, sex, or clothing have secondary survival value. Loss of minimal levels of these resources is critical.

Looking at more solid empirical data, Tversky and Kahneman (1981) showed compelling evidence that individuals overweight the consequences of loss in their decision making and that this is done without awareness of this tendency. Similarly, Tymstra (1989) used the term "anticipated decision regret" in regard to decision making for medical patients faced with new technologies. He finds that in order to prevent the chance of a loss, people sacrifice resources that are vastly disproportionate to the probability of the loss occurring. So, for example, women will submit themselves to dangerous, time-consuming medical procedures that are unlikely to result in their ability to achieve fertilization in order to avoid some future moment of regret when their theoretical loss is felt (e.g., "I might have had a baby and now I do not").

Applied to burnout, COR theory suggests that workers are more sensitive to workplace phenomena that translate to losses for them. For teachers, negative interactions with parents, problem children, and negative evaluations by administrators will be more salient than everyday rewards they might receive. Shamir (1986) noted that the aggregate of all the negative aspects of the job join to produce burnout. Losses on the interpersonal level in the form of interpersonal conflict appear to be particularly salient in the burnout phenomenon and may represent one of the ongoing areas of stress that takes a daily toll on workers (Leiter & Maslach, 1988). Further, Burke (1989a) argues that the major contemporary sources of stress in the business world are loss-related. They include mergers and acquisitions (which decrease number of management jobs), organizational retrenchment and decline (which lead to staff and resource cutbacks), and insecurity of the job's future (i.e., potential for future loss).

Secondary Importance of Gain

Although gain of resources is less important than prevention of loss, it is by no means inconsequential. Gain is important on two grounds. First, gain has indirect value because possession of resources decreases the chance of loss. Resources reduce vulnerability both to immediate loss following some event or circumstance and to the ensuing loss that follows in the wake of initial loss. By possessing high self-esteem, for example, individuals are less shaken by job layoffs and in turn are less likely to have marital problems that ensue with the likelihood of further loss of self-esteem during a prolonged job search. Similarly, those rich in resources are less likely to encounter stressful events, as their resources distance them from many of the high-risk circumstances that accompany the lack of resources. Money, insurance, loved ones, and job security are examples of resources, and having them places people in a less vulnerable position.

Second, in addition to these indirect advantages, there are direct positive effects of resource gain. In this regard, we strive to gain resources because their possession is valued in itself, in that social praise and status, as well as personal feelings of success, are consequent to possession of strong resources.

Lack of gains, however, should not be as powerful a predictor of burnout as is loss. In this regard, unmet expectations (i.e., expectations of gains on the job) were not found to predict burnout. In contrast, role conflict—which produces direct losses and lost opportunity—was strongly related to emotional exhaustion, a key facet of burnout (Jackson, Schwab, & Schuler, 1986).

Investing Resources to Prevent Loss

According to the COR model, what individuals have to offset resource loss is other resources. For example, money may be invested to increase knowledge and contacts (e.g., attending Harvard), time and energy may be invested to improve the quality of one's marriage, and self-esteem may be invested in order to make further gains in esteem or to tackle a work problem and gain status.

The term *investment* is chosen because "used" implies "used up," and this would be inaccurate. Many resources are employed without diminishment, but, like investments, when employed they are placed at a certain risk. Let us use an example. A senior executive forcefully forecasts a market trend against the judgment of her colleagues. She has, in a sense, invested her esteem in the eyes of others, and future market behavior will probably be watched to see whether or not she was correct. If she was correct, and her opinion was outspoken, she will gain kudos, and her esteem will have risen. If, on the other hand, her predictions are not borne out, then her esteem as judged by others may drop. This, in turn, will affect the way she feels about herself. In this way, it can be seen how a very nontangible personal resource may be invested, along with the consequences for loss and gain of that resource.

One clear example of how lack of gain following investment can contribute to work stress and burnout is provided by Burke (1989a). He argues that "locking in" is one of the key stressors at work. Locking in occurs when a worker is unable to move further up the job ladder despite hard work and initiative. Since there is no loss or threat involved, the only aspect of this circumstance that is stressful is the failure to gain following investment of resources such as time, skills, and education. It is possible that because teachers, nurses, and social workers have limited upward mobility, they are especially vulnerable to burnout.

Evidence of the Primacy of Loss

There have been a number of approaches to understanding what element is basic to stress. COR theory proposes that loss is primary, but let us consider some of the other leading approaches.

Stressfulness of Transition and Change

Another avenue of thinking has suggested that stress ensues from transition or change per se. However, a careful examination of what is now a wealth of research findings suggests otherwise. An excellent review by Thoits (1983) indicated that positive events not only are not stressful but, on the contrary, they fill a potential stress-buffering role. Cohen and Hoberman (1983), for example, found that those who experienced positive events were less susceptible to the negative effects of undesirable events.

Some confusion in this regard is related to the fact that change is implied in the wording of many items on stressful events measures, such that responses reflect a mix of negative and positive change. However, when negative change is removed, the remaining effect of positive change is shown not to be detrimental. Confusion on this issue has also followed from reports that workplace relocation is stressful. However, such relocation often involves loss of resources. In a recent review of the literature, Munton and Forster (1990) support this point. Job relocation involves a simultaneous loss of social networks, friends and relatives, social roles and membership, and involvements that have been created to ensure a sense of meaning and esteem. This loss is greatest for spouses and children. So it is not surprising that there is little indication of deleterious effects on employees who are transferred following promotion (Latack, 1984). They gain in many resources and reenter a workplace with a web of built-in resources. In contrast, their wives (typically men are transferred) experience increased rates of depression and their children encounter serious school problems. Similarly, retirement was not found to be stressful unless retirees had inadequate finances or poor health (see Kasl, 1980; McGoldrick & Cooper, 1985).

Such arguments may be taken to mean that transitions and change are stressful. However, the point is that it is not the transition or change that is stressful but rather the loss that accompanies the transition. Positive change means resource gain and decreased vulnerability to loss, and, as opposed to being stressful, it is beneficial and healthful.

Role Ambiguity

Still another line of thought speaks to the issue of role ambiguity as a work-related stressor (Burke, 1989a). Here again, ambiguity needs to be unpacked. It is questionable whether an ambiguous situation where rewards were constant, such as the unconditional positive regard of Rogerian therapy, would prove stressful. Work roles are likely to be ambiguous when supervisors wish to retain power or when it is not clear what should be done to obtain goals. In both cases, people are forced to invest resources without any guarantee of payoff. This investment without expectation of gain would be stressful according to COR theory (see Chapter 4). COR suggests that people plan strategies to obtain goals. This leads to both a need for a degree of control and a clear playing field on which to carry out the strategy. Ambiguous work situations frustrate these goal-directed efforts.

Preliminary Empirical Support for Loss's Primacy

The above arguments may be seen as convincing or not, but they are open to criticism in that they are not direct examinations carried out specifically to study loss versus gain. In order to meet this criticism, we gave two samples questionnaires on two occasions separated by approximately 3 weeks. Two hundred and fifty-five students were included in one sample and 74 community residents in the other sample. Participants rated their losses and gains from a comprehensive list that was devised to study COR theory (see Hobfoll, Lilly, & Jackson, 1992). They also responded as to their state and trait anxiety and to their depressive mood. Losses were found to have a deleterious effect on each measure of psychological distress. In addition, for both samples, gains had a zero-order effect. The only evidence to the contrary is that greater gains have some modest association with lower distress.

COR theory further postulates that gains are important after losses are taken into consideration. This was examined via hierarchical regression analysis by which gains were entered in the regression equation following the entry of losses. When this procedure was performed, gains were seen as having a marked impact on psychological distress. However, as predicted by COR theory and in contrast to change or transition theories of stress, the effects of gains were positive, not deleterious. This moderator effect was noted for both samples on each of the three measures of psychological distress. It is notable that the moderator effect was marked in each case.

These data clearly highlight the primacy of loss and the secondary, albeit important, nature of gains. It is worth noting that the data were obtained on separate samples, at two points in time, using a variety of psychological measures. These factors bolster our argument and complement the indirect evidence we have already noted from other studies.

Losses Cue Coping Responses

COR theory suggests that when loss occurs people engage in coping more actively in order to reduce the effects of loss. This process was studied by Freedy, Shaw, Jarrell, and Masters (1992) in an examination of hurricane victims in Charleston, South Carolina. Staff and faculty at a school of medicine (*N*

= 418) were administered questionnaires on which they indicated their degree of resource loss following the devastating effects of Hurricane Hugo. As predicted, the greater the loss experienced, the more coping behavior followed. For some respondents the coping was of the more effective problem-solving type; for others the coping behavior was the usually less effective use of denial and avoidance. What we wish to emphasize here is that what was common for both groups was the increase of whatever coping behavior that they chose to mobilize. Also, the process of withdrawal can be seen as an effort to limit loss, similar to the process of social distancing others noted in the burnout syndrome (Cherniss, 1980a). Usually viewed as a negative response, such coping behavior may actually enhance people's ability to otherwise stay engaged in the stressful situation.

This coping also comes at a cost. As resources are invested, there is a likelihood of resource depletion. In the case of social support, for instance, individuals can call on social resources, but when the need is chronic, the future availability of support decreases. Dunkel-Schetter (1984) noted in this regard that more social support was offered to individuals who coped effectively— paradoxically, those who probably needed less support. In contrast, the more severely decapacitated group was less likely to obtain continued help. Hobfoll and Lerman (1988, 1989) also found that under conditions of chronic stress, both the availability and the positive effect of social support decline. Indeed, those who were exposed to chronic stressful conditions were found to be burdened, not aided, by support after a year of exposure to the chronic stressor. Likewise for burnout, the experience of chronic stress is likely to deplete social resources and leave individuals increasingly vulnerable (Shirom, 1989).

This cost of coping helps explain why possessing an array of strong resources aids stress resistance. Resources are related to one another, and as one resource is depleted another may substitute. In fact, one resource may limit or prevent depletion of a second resource. Money can limit loss of time by purchasing services, and insurance can prevent a loss of money following an accident or illness. Positive self-regard increases both one's social desirability and the ability to employ social resources during times of need (Hansson, Jones, & Carpenter, 1984; Hobfoll & Lieberman, 1987).

The weblike nature of resources also suggests that resource loss and gain will occur in spirals. Loss spirals will follow initial losses, with each loss resulting in a depletion of resources for confronting the next threat or loss. Resource impoverishment also prevents institution of gain cycles that might reverse loss spirals. Similarly, a gain in resources emboldens individuals to attempt further investment and take greater risks, now with the resources to reduce the possibility of loss. For example, success on a task will increase a worker's sense of mastery. In turn, the individual will be likely to take on harder tasks, especially if provided added resources that are related to task accomplishment.

Other stress theories are mainly reactive in that they predict responses to stressful circumstances (Lazarus & Folkman, 1984; Selye, 1951-1956). Being a motivational model, COR theory makes predictions of behavior during non-stressful circumstances as well as stressful ones. Specifically, the desire to avoid loss is continually present, pressing individuals to seek resource gains in order to avoid the possibility or severe consequences of future loss. The goal here is to create sustained gain cycles. Employees would thus be seen as working on the stress in their lives even when no particular stress was at hand. Favors are provided, contacts made, and paperclips stocked in order to prepare for future threats or losses. Because all individuals have experienced loss in the past, they learn to prepare for it in the future. Seen another way, workers are engaged in defending against burnout in the day-to-day strategy of their approach to work, especially if resources for adjustment are available to them.

Failure to Gain Resources Following Investment

According to most stress theories, stress occurs when people are challenged or threatened beyond their coping capacity (see Lazarus & Folkman, 1984). This framework skirts the problem of why individuals are often under stress when there is no apparent problem at hand. In this regard, we are not referring to neurotic processes but to the very common problem found among people who feel that they are just treading water. According to COR theory, one of the three principal mechanisms by which stress occurs is when people invest resources without consequent payoff.

By investing resources there is, in a sense, a buildup of losses. In this regard, coping with everyday challenges is translated into costs of time, energy, other missed opportunities, trust, and other valued resources (Schönpflug, 1985). Without consequent gain, this registers as a net loss of resources, which is stressful. Furthermore, resources are invested in order to obtain other resources. For example, workers invest their time for the salary that will afford them a reasonable lifestyle, for job security, for retirement security, for status, and for a sense of accomplishment. When these goals are not achieved, both the invested resources and the expected gains are lost. These resource goals are imagined and these cognitions are accompanied by deep feelings. The desired sailboat is linked with fantasies that already include the feelings of relaxation and satisfaction that will be derived as well. Retirement is imagined as a time free of financial worries and full of fishing trips and grandchildren. When these payoffs are not forthcoming, the loss of resource goals is experienced as stressful.

The concept of failure to gain resources following investment is critical to understanding occupational stress. Most people expect a promised or assumed payoff for their work, and yet circumstances often occur to frustrate those expectations. Companies merge, younger people take one's place, pensions are mismanaged, and housing prices increase so that the dream house may never be attained. On a day-to-day level, the failure of an organization to accept input of experienced workers, the degrading way supervision is sometimes conducted, inflation, and lack of acknowledgement for special efforts take their toll (Burke, 1989a). This bitter pill is often accentuated for women and minorities who witness the differences in the rewards they gain compared to white male counterparts—since the disparity between effort and gain is facilitated by close social comparisons (Eckenrode & Gore, 1990).

Hassles and Loss

In any discussion of stress it is impossible to avoid the issue of hassles and the ongoing argument of whether hassles or major events are more closely related to stress (Dohrenwend, Dohrenwend, Dodson, & Shrout, 1984; Lazarus, De-Longis, Folkman, & Gruen, 1985). Many human service positions in particular are fraught with repetitive hassles. COR theory approaches this issue from a different perspective than has typically been taken. According to COR, all events may be stressful to the extent that they result in resource loss, threat of loss, or failure of gain on investment. Hassles are likely to result in resource loss on a number of grounds and, to the extent that they do not, they should not be stressful. One exception to this was raised earlier: specifically, any event that physiologically leads to exhaustion will be stressful for physiological reasons. No psychological stress theory is necessary to understand this, but it is surprising how often this point is overlooked. One recalls Charlie Chaplin's classic film *Modern Times*, in which Chaplin, as an assembly line worker engaged in a repetitive task, leaves the line with a dazed expression, unable to prevent his body from continuing to turn the now imaginary screws and bolts.

Reviewing lists of hassles (Kanner, Coyne, Schaefer, & Lazarus, 1981), it is obvious that most are related to loss. Empirical evidence clearly demonstrates that the most disturbing hassles are tied to a central loss event, such as care of an elderly parent. Indeed, Stephens, Norris, Kinney, Ritchie, and Grotz (1988) noted that even these hassles are not terribly harmful if the relationship remains intact. However, they found that when dementia sets in, the everyday hassles of providing care take a much more severe toll. Most hassles, if occasional, are tolerable. For many human service workers and managers, daily hassles are tied to a central loss of professional effectiveness or of their ability to actually perform the work for which they were trained (Landsbergis, 1988; Sexton, 1982). As in the Stephens et al. study, it is not the small hassles but the symbolic loss they represent that is critical.

Examination of the workplace reveals that many related hassles do indeed involve loss of resources. At work, minor bureaucrats keep people from achieving goals, endless paper work represents loss of doing the nursing work that nurses want to do, and unthankful patients represent a loss of expected rewards central to the Florence Nightingale fantasy (Sexton, 1982). Traffic prevents people from being where they want to be and achieving the gains associated with work or family. Concert music on the car stereo can limit the physiological drain, and air conditioning can limit the stress caused by heat and unpleasant exposure to noxious fumes, but the time still seems to be invested without payoff. In a study of nurses, for example, it was found that negative interactions with coworkers were particularly stressful (Leiter & Maslach, 1988). These negative interactions were not common, but they give the impression that they picked away at nurses' workplace satisfaction.

This begs the question of why we need to appeal to an overarching theory when we can simply list the hassles that represent a lower level of abstraction. Without an overarching theory, we are left with the need for endless lists of hassles for which we can only know post facto which are the most stressful. COR turns this around and questions the degree of loss, threat of loss, or failure of gain. This can be objectively derived or subjectively assessed, depending on the context and the nature of the problem raised. For individuals, subjective assessments are critical; when intervening or understanding organizations or systems, objective measures may be more helpful in unpacking the causes of stress and how to alleviate them.

COR THEORY AND INTERVENTION

COR theory suggests that intervention should be based on enhancing resources and eliminating vulnerability to resource loss. It implies that perceptions should be seen as real rather than as products of personality differences. If you will, cognitive approaches have painted a picture that suggests that people view reality with different cameras. In other words, it is the individual differences in perception that are underscored. COR theory argues instead that people have similar cameras, with similar lenses, and that only the focus is somewhat individual. Greater emphasis is thereby placed on objective factors such as the underlying events and circumstances that shape perceptions, the latter being seen as approximate catalogues of reality.

A cognitive approach to stress management might act to enhance people's sense of personality hardiness (Kobasa, 1979). COR would suggest, however, that hardiness is based on possession of many resources and a history of successful coping owing to this surplus. A recent study by Ozer and Bandura (1990) emphasizes the importance of actual resources and how they underlie efficacy cognitions. In this study, women participated in an empowerment program in which they mastered the physical skills required to defend themselves against unarmed sexual assailants. Participants experienced enhanced perceived coping and cognitive control efficacy, decreased perceived vulnerability to assault, and reductions in the incidence of intrusive negative thinking and anxiety arousal. Behaviorally, women experienced increased freedom of action and decreased avoidant behavior. It may also be added that in the event of an assault these women were actually more capable of defending themselves than women who had not participated in such a program, thus minimizing traumatic losses associated with being assaulted. Forty-three women were involved in the controlled study, but many more have participated in the training program. Of the hundreds of women who have completed this program, 40 have reported being assaulted. Of these, 38 escaped rape—30 stunned and disabled their assailants, and 8 scared the assailant away. The two raped women were attacked by an armed assailant. Clearly, selective reporting may play a role in this aspect of the data; nonetheless, the results suggest a process whereby resources are translated to more successful coping in a behavioral arena, and they match the more methodologically sound aspects of the controlled study.

A more strictly cognitive approach might choose desensitization to the anxiety involving exposure to possible assault. Translating this to workplace stress, many interventions aim to get target groups to perceive stressors differently. Such an approach is faulty on a number of accounts. First, assessment of coping resources is a better predictor of behavior than anxiety. In this regard, people tend to avoid situations and activities that they perceive as exceeding their coping capacity, choosing instead to engage with environments and challenges that they believe are within their range of capabilities (Bandura, 1989; Betz & Hackett, 1986). As others have also pointed out, based on their competencies (i.e., actual resources), people affect their environment in a way that tends to reify their self-beliefs (Coyne, 1976; Swann & Brown, 1990). Persons suffering from depression, for example, believe that they are interpersonally unskilled and thus interact with others in ways that increase the likelihood of their receiving negative social feedback (Coyne, 1976). In contrast, those with social skills are likely to successfully engage their environment and receive more positive feedback from others (Hobfoll & Lieberman, 1987; Hobfoll & Lerman, 1989). Intervention efforts must focus first on the resources that underlie perceptions, with an expectation that perceptions will change with changed circumstances.

Workplace Intervention

Recent work by Leiter (1990a) suggests how COR theory might be applied to workplace stress. Using structural modeling, he found that family resources, worksetting resources, and coping style were independently related to different aspects of Maslach's (Maslach & Jackson, 1986) burnout construct. Consistent with COR theory, the resources were related to initial levels of burnout and further burnout over time. This points to a causal process whereby those possessing greater resources are not only less likely to experience burnout but are more likely to recover if they encounter burnout.

Leiter emphasizes that family, work, and personal coping resources are of equal importance in the management of occupational stress. The boundaries between work and family are seen in this way as somewhat arbitrary. These conclusions accentuate the need to address resources that lie in the individual and in the major life settings of importance to the individual. Workplace interventions should therefore approach the problem of occupational stress as involving both home and work spheres. These conclusions are echoed in the recent work of Greenglass and Burke (1988), Orthner and Pittman (1986), and Jackson and Maslach (1982).

In contrast to the strategy outlined above, there has been a tendency to deal with stress in the workplace both superficially and on the symptom level. We have reviewed stress management programs at numerous worksites and found that typically they are one-shot workshops. A second type of program attempts to alleviate stress through relaxation and biofeedback. Because all stress theories suggest that stress has real causes (even if individuals may exacerbate the problem), these attempts seem destined to failure. At best, they may alleviate some symptoms, but at the same time they are likely to lead to greater stressors because the causative factors are tied up in loss cycles. Despite numerous theories that support our thinking, interventions overwhelmingly avoid the causes of stress and focus on the symptoms of the victims (Singer, Neal, & Schwartz, 1989; Rosch & Pelletier, 1989, for support of this point).

COR theory, as outlined above, can be applied to worksite interventions, leading to a very different kind of program, but one that is similar to Leiter's suggestions. Table 7-1 presents an overview of the key aspects of a worksite stress reduction program aimed at reducing workplace burnout. Such interventions should focus first on causes of losses. They should tap the subjective nature of the stress experience and the objective aspects as well. Three foci of intervention are critical: (1) building and enhancing resources, (2) interrupting loss chains, and (3) activating gain spirals.

First and foremost, COR theory concentrates on the causes of losses. The most important of these occur on the system level. When so many nurses, or desk sergeants, or air traffic controllers claim that their jobs are stressful, this is probably the case. Too often psychologists turn to individual difference variables or differences in perceptions when confronting stress when they probably should concentrate on the objective causes of worksite stress. Management encourages and perhaps forces this kind of thinking (Singer et al., 1989). Changing the system challenges management, whereas changing individuals fits the management's model of "if only workers were better, things would be OK." Because psychologists and other organizational people are hired by management, there exists at this juncture a potential threat to the integrity of the intervention.

Some causes of loss are inevitable. For example, frequent loss of patient life in intensive care cannot be prevented. However, many aspects of workplace stress are related to how the system is designed. Intervention on the level of the

TABLE 7-1   Key Aspects of a Worksite Stress Reduction Program

---

1. Emphasis on causes of losses
   A. Systems level (primary)
   B. Individual level (secondary)
2. Focus on objective and subjective nature of stress
   A. Objective (primary)
   B. Subjective (secondary)
3. Focus on building resources and enhancing availability of resources
4. Interruption of loss chain
   A. Rooting out initial sources of loss (primary)
   B. Changing more advanced stages in loss chains (secondary)
5. Activation of gain spirals
   A. Enhancing environments that facilitate gain
   B. Increasing individual gain efforts

---

group or organization can do much to decrease the probability of loss, to cut off loss cycles, and to enhance the focus and thus the potential for gain.

In a recent study, Freedy (1990) endeavored to incorporate the principles of COR theory in a workplace intervention for nurses. Six area hospitals were chosen and groups were placed in a waiting list design. Nurses received either a dual-resource, mastery– and social support–enhancing intervention; or a single-resource, mastery-enhancing intervention. Groups met on five occasions. It was predicted that the dual-resource intervention would be more beneficial because both resources are ecologically required for successful coping of nurses. In addition, the social support aspects of the program focused on both workplace and home support. This latter aspect of home support was seen as critical because nurses (only women in this sample) are often most challenged by the interface between work and home-related stressors. The groups were matched for time of intervention, discussion, and directiveness of the groups. The nurses were encouraged to work together on group and system level problems that were not amenable to change via individual efforts.

Consistent with the project design, the resource of social support was enhanced in the dual-resource intervention but not the single-resource intervention. The enhancement of social support occurred across both work and home environments. This treatment effect persisted when evaluated at a 5-week follow-up. Results concerning the enhancement of mastery were more mixed. Level of mastery did indicate signs of enhancement, but initial positive effects dissipated during the follow-up period. Gains in mastery were more likely to occur in the dual-resource intervention. Although predicted, this effect may appear counterintuitive because participants in the single-resource mastery-only intervention received a greater dose of mastery-focused intervention. Also consistent with COR theory, the enhancement of social support and mastery resources was related to clinically significant decreases in depression scores for nurses in the dual-resource intervention.

It appears that a five-session program was only meeting a minimum threshold for change. Also, although Freedy encouraged work on system levels, the nurses may not have been or felt empowered to implement certain important changes. Furthermore, many women suggested that their spouses needed to participate in the program. More specifically, spouses should be targeted for intervention pertaining to the home workload because this responsibility disproportionately falls on the nurses at home as wives and mothers. These criticisms are consistent with our own thinking, and we agree that intervention needs to be even more broad-based in terms of both time and the players involved.

CONCLUSIONS

Conservation of resources theory may serve as a valuable heuristic for burnout research. It emphasizes the real-world aspects of stress on both the system and the individual level, highlights that people need resources to meet challenges successfully, and offers a framework that allows for the simultaneous study of process and outcome. It is clearly an integrative conceptual schema that belies attempts to separate domains of work and family, individual and group, or input and output of efforts. In this way the concept of resources joins these different domains as a common economy in which resources are exchanged. It follows that research aimed at further understanding occupational stress and the interventions designed to alleviate this stress, and consequent burnout, should target those aspects of the process that deplete resources, minimize gain of resources, devalue obtained resources, and facilitate resource loss (see Shirom, 1989). Perceptions will always be important, but when repeated studies indicate that professions and occupations share common concerns, one must ask why there is always a turn to the individual. This victim-blaming approach is certainly easier, and it is to management's short-term benefit; however, it is difficult to see its meaningful long-term impact. Ongoing losses will eventually dissolve the sense of calm or mastery in anyone but a Zen monk. Reality has too rough an edge and occupational stressors typically repeat themselves either continuously or in cycles, wearing away at any superficial remediation.

Hopefully, researchers, interventionists, union representatives, and management will focus their attention on cycles of loss, break these loss chains, and maximize the opportunities for resource gain. Much needs to be learned about the value and nature of particular sources of resources, but it should not be understated that we already know a great deal about resources that are valued in our culture and that can be used to reverse or prevent burnout.

III

ORGANIZATIONAL
APPROACHES

After the interpersonal and individual approaches to burnout, four organizational approaches are discussed in this part of the book. The organizational environment is essential in understanding burnout because it is defined as a negative, *work*-related psychological phenomenon. The approaches that are described below consider burnout as negative organizational behavior, which not only affects the individual but the organization as well. Although the relevance of the organizational environment is self-evident in the case of burnout, remarkably few approaches exist that provide a theoretical framework for understanding burnout in an organizational context. Each of the four approaches included in this part of the book tries to fill this gap from a particular perspective.

Cherniss starts by arguing that the notion of self-efficacy can serve as a unifying conception to integrate most writing and research on burnout. Self-efficacy refers to a sense of mastery and the belief that one can exercise control over events that affect one's life. Successfully and independently achieving one's goals enhances self-efficacy. Failure to achieve these goals leads to psychological failure, decreased self-efficacy, and eventually burnout. In this regard, Cherniss's position converges with those of several other contributors to this volume (Pines, Hallsten, Hobfoll and Freedy, and Burisch) and is more "individual" than "organizational." However, Cherniss goes on to argue that self-efficacy is not a personality trait but is linked to a particular role. For this reason he introduces the concept of *professional* self-efficacy: the belief that one is able to perform well in professional work roles. Cherniss distinguishes three domains of professional self-efficacy or role performance: task, interpersonal, and organizational. The latter domain is especially important, as demonstrated by Cherniss's research and illustrated by a case study. Finally, Cherniss considers four factors that influence professionals' job performance: internal performance standards, knowledge and skill in each of the domains of professional self-efficacy, predictability and controllability of the work environment, and the social status accorded to the professional.

Winnubst analyzes the relationship between organizational structure, social support, and burnout. He argues that particular organizational structures produce particular organizational cultures. For instance, a machine bureaucracy that is characterized by standardization of work and formalization tends to reinforce perfectionism and conformity. On the other hand, a professional bureaucracy, that is characterized by standardization of skill and little formalization, tends to reinforce creativity and autonomy. According to Winnubst, social support systems are closely linked to the organizational regime. In a machine bureaucracy, hierarchy and authority play an important role; therefore most communication is vertical. Social support is mostly instrumental and aimed at maintaining predictability. In a professional bureaucracy, team work and guidance are more prominent principles. Accordingly, communication is horizontal as well as vertical. Social support in professional bureaucracies is informational and emotional. Winnubst argues that the antecedents of burnout differ depending on the organizational structure and the institutionalization of social support. In a machine bureaucracy, burnout is caused by emotional drain due to routine, monotony, and lack of control. In a professional bureaucracy, burnout is caused by the relatively loose structure and the resulting continuous struggle with other people (ie, role problems, domain fights).

In their contribution, Noworol, Żarczyński, Fafrowicz and Marek argue that burnout blocks creative behavior in organizations. According to adaptation-innovation theory, two types of employees exist. "Adaptors," who behave rather conventionally, come up with agreed problem definitions and accepted solutions. They are neither original nor creative. They are concerned with "doing better" rather than "doing differently." In contrast, "innovators" redefine problems and prefer less acceptable and more original solutions. They are more creative and concerned with "doing differently" rather than "doing better." The authors elaborate on the concept of creative style of behavior, which has three domains, each of which is divided into several subdomains. Their central thesis that burnout is associated with a poor creativity style of behavior is supported for almost all domains in a sample of Polish managers. Moreover, as hypothesized, the majority of the burned-out managers could be characterized as adaptors instead of innovators.

In the final contribution, Cox, Kuk, and Leiter try to integrate burnout in a more comprehensible transactional model of occupational stress, which has been developed by the first author. Most importantly, stress and burnout are discussed in relation to organizational healthiness. In this transactional stress model, being "worn out" (i.e., tired, emotionally labile, and cognitively confused) and feeling "uptight and tense" (ie, fearful, tense, and anxious) are two indicators of a poor state of overall well-being. The authors demonstrate that emotional exhaustion (which they consider to be the central dimension of burnout) is theoretically as well as empirically related to being worn out. Depersonalization is regarded as a coping strategy, whereas the third component of burnout (personal accomplishment) is viewed as an appraisal outcome. The authors conclude that burnout is a particular slice across the transactional stress process that is common in human service workers. In the next step they try to explain why only relatively weak correlations have been found between stress and burnout. The authors argue that organizational healthiness affects the experience of stress as well as employee health. Therefore, organizational healthiness has to be included as a moderator variable in the stress–burnout relationship. Cox, Kuk, and Leiter provide compelling empirical evidence for the moderating role of organizational healthiness, in terms of teachers' feelings of being worn out. In addition, a direct effect of organizational healthiness on feeling worn out was observed. A likely hypothesis is that the same results will be obtained with burnout.

The four approaches in this part of the book offer interesting perspectives for future research as well as for relating burnout conceptually to professional self-efficacy (Cherniss), organizational theory (Winnubst), creative behavior in organizations (Noworol et al.), and organizational healthiness (Cox, Kuk, and Leiter).

8

ROLE OF PROFESSIONAL SELF-EFFICACY IN THE ETIOLOGY AND AMELIORATION OF BURNOUT

Cary Cherniss
Rutgers University

In reviewing everything we have learned about professional burnout during the last 15 years, is there a single, underlying theme that emerges? Given how large and varied the research literature on burnout has become, it's not clear that one theme can tie it all together. Yet the very diversity of the work demands that we make some attempt to develop a unifying conception. A particularly good candidate for such a unifying conception is the concept of self-efficacy. It is not a new concept in psychology. But most writing and research on burnout has not explicitly recognized this conceptual link between burnout and self-efficacy. This is unfortunate, because linking burnout with self-efficacy can point to some valuable new directions for research, theory, and action on burnout in the future, as Leiter (1990a) recently suggested.

When I first began to study professional burnout in the mid-1970s, I was not aware of Bandura's (1977, 1982) work on self-efficacy. And Maslach (1982a) had not yet isolated "personal accomplishment" as an important dimension of burnout. However, I was familiar with the work of Robert White (1959), who had argued that competence is a strong human motive. And in my own research on new professionals, I found that when this motive was thwarted, the result was high levels of stress and in some cases burnout. My own subjects seemed to be telling me in various ways that achieving a sense of competence in their work was a particularly important concern. It seemed to lie behind so many of the overt "sources of stress" to which subjects referred. For instance, when professionals complained about clients who were resistant or did not improve, the professionals seemed to be distressed primarily because these client behaviors prevented the professionals from feeling competent and successful.

Similarly, when my professional subjects complained about excessive workloads, lack of administrative support, and bureaucratic constraints, the underlying issue again seemed to be that they could not feel successful and competent–not because they lacked the skill and ability, but because systemic factors prevented them from using those skills in a way that would achieve intended outcomes.

My original work involved novice professionals, so one might argue that it was not surprising that achieving a sense of competence was such an over-riding concern. However, when I studied this same group over 10 years after they began their careers, I found that achieving a sense of competence still seemed to be of great importance in determining how they felt about their work (Cherniss, 1989a, 1992). Most of the subjects felt considerably more confident and sure of themselves at this point in their careers than they did when they first began practicing. However, many continued to feel that they could not perform as well as they thought they should. Because it is extremely difficult for an experienced professional to admit to feelings of inadequacy, these subjects did not often come right out and express their concerns. The continued importance of competence sometimes was revealed only when the professionals had to make a significant change in their work roles and had to perform in new ways.

For instance, one special education teacher talked about how upset she was when she was involuntarily transferred from a resource teacher position to teaching in a self-contained classroom. This teacher had more than 4 years of experience at this point, and she was seen as a respected and competent professional. But as she spoke about this experience, it became clear that a major source of her upset was fear that she would not be able to perform competently in the new role. Here is a revealing excerpt from my interview with her:

> *I was transferred, involuntarily, into a situation where I had a self-contained class. And that was very upsetting to me. I didn't have a choice about it. . . . I was real frustrated about the whole thing. And I think mostly I felt like, "I'm not*

sure I can do this." I was scared about it. It's a different type of student to deal with on the full-time. So I was real uncertain about it.

Fortunately, she was able to find another teacher who had taught in this kind of setting for several years and who was willing to help her learn the new skills that she needed to perform competently in the new assignment.

BURNOUT AND SELF-EFFICACY: LEWIN AND HALL'S WORK

This link between self-efficacy and burnout was reinforced for me when I read Hall's (1976) work on psychological success. Building on the earlier work of Lewin (1936), Hall proposed that work motivation and satisfaction were enhanced when a person successfully and independently achieved a goal that was challenging and personally meaningful (see Figure 8-1). Such achievement led to "psychological success," which in turn encouraged the individual to become more involved in the job, to set more challenging goals, and to feel more self-

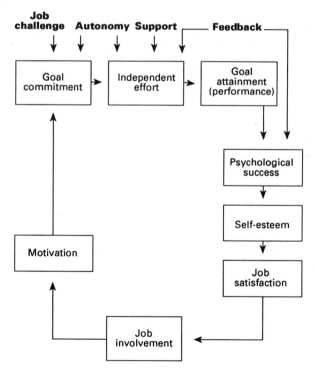

FIGURE 8-1 Hall's model of psychological success. Adapted from Hall (1971, p. 66).

esteem. Hall stressed that what was important was a person's feelings of success rather than success as objectively measured.

Particularly interesting from the perspective of burnout research and theory were Hall's ideas about what would happen if an individual were not able to experience psychological success. Influenced by Argyris (1964), Hall wrote that the person would "psychologically withdraw from those arenas in which he or she is experiencing failure" (Hall, 1976, p. 136). Psychological withdrawal, of course, is a major characteristic associated with burnout. More specifically, Hall (1976, p. 191) proposed that "psychological failure" would lead to a person's:

1. Withdrawing emotionally from the work situation by lowering one's work standards and becoming apathetic and disinterested
2. Placing increased value on material rewards and depreciating the value of human or intrinsic rewards
3. Defending the self-concept through the use of defense mechanisms
4. Fighting the organization
5. Leaving the organization

These were precisely the changes I observed in my first study of professional burnout (Cherniss, 1980a). They also resemble some of the changes Maslach (1982a) associated with the depersonalization dimension of burn-out.

In defining the concepts of psychological success and failure, Hall not only described the symptoms of what later became known as burnout; he also identified the job and organizational characteristics that tend to be associated with lower levels of burnout in the research literature: challenge, autonomy and control, feedback of results, and support from supervisor and coworkers (Drory & Shamir, 1988; Leiter & Maslach, 1988; Maslach & Jackson, 1984a; O'Driscoll & Schubert, 1988; Pines & Aronson, 1988).

Thus, Hall's work strongly reinforced my sense that there was a link between burnout and an inability to achieve a sense of competence or success in one's work. Factors in the individual or the work situation that enhanced feelings of success and competence would reduce burnout, while factors that promoted feelings of inadequacy and failure would increase burnout.

THE LINK BETWEEN SELF-EFFICACY AND BURNOUT

Bandura's thinking about self-efficacy had not been widely disseminated when Hall was developing his ideas about psychological success and failure in work settings. However, it is consistent with Hall's thinking and with much of the research on burnout that has emerged over the past 10 years.

Bandura defines perceived self-efficacy as "people's beliefs about their capabilities to exercise control over events that affect their lives" (Bandura,

1989, p. 1175). In a recent review article, he cited numerous studies showing that stronger perceived self-efficacy leads to the setting of more ambitious goals and a firmer commitment to these goals. Stronger self-efficacy also leads to more effort and persistence in pursuing goals.

In addition to linking self-efficacy to commitment and motivation, research also suggests strong links between self-efficacy and stress. People with stronger perceived self-efficacy experience less stress in threatening or taxing situations, and situations are less stressful when people believe that they can cope successfully with them (Bandura, 1989). Because burnout is typically regarded as a reaction to adverse, stressful situations, this relationship between self-efficacy and stress suggests a link between self-efficacy and burnout as well.

Some critics of Bandura's work have argued that as a cognitive concept self-efficacy puts too much emphasis on the person's responsibility for maladaptive behavior, and too little on the environment's. This criticism is especially important for those of us who have done research on burnout because the question of personal versus environmental causation has been a central one. To date, most of the research tends to suggest that adverse organizational conditions are more significant in the etiology of burnout than are personality factors (Gerstein, Topp, & Correll, 1987; Pines & Aronson, 1988). Thus, focusing on a concept like self-efficacy would seem to be inconsistent with prevailing views about the causes of burnout.

Bandura recognizes that there are environments that are so unresponsive, unjust, and punitive that strong self-efficacy by itself is not sufficient for positive adaptation. But he also argues that people's responses to such environments will differ in important ways depending on the strength of their self-efficacy. Those who perceive themselves to be more efficacious will engage in social activism; and, if their efforts to change the environment meet with repeated failure, they will eventually look for better environments in which to work. But those who are low in self-efficacy will tend to react to unresponsive environments with apathy, resignation, and cynicism. Thus, strong self-efficacy ultimately promotes environmental change as well as individual adaptation.

Furthermore, Bandura and others recognized that environmental factors influence a person's self-efficacy. An individual in an environment that undermines self-efficacy will not feel as efficacious. Thus, the dichotomy that we have drawn between personal and environmental causation ultimately may prove to be an illusory one; self-efficacy seems to involve a reciprocal relationship between personality and environment. Thus, in giving self-efficacy a central etiological role in burnout, we need not minimize the contribution of environmental conditions. In fact, self-efficacy theory helps explain why certain job characteristics, such as low autonomy, are especially conducive to burnout.

SELF-EFFICACY AND MORE RECENT VIEWS OF BURNOUT

More recent theoretical ideas about the burnout process also are compatible with the notion that professional self-efficacy plays a central role. For instance, in Hobfoll's conservation of resources theory (Chapter 7), mastery is identified as one of the basic resources for adaptation (along with social support and self-esteem). Without a sense of mastery, i.e., self-efficacy, there can be little chance of adaptation, and burnout becomes more likely.

Hallsten (Chapter 6) also points to the central importance of self-efficacy when he offers a nice distinction between alienation and burnout. According to Hallsten, alienation is likely to occur when the individual has little freedom in accomplishing a possible task whereas burnout is likely to occur when the individual has much freedom in accomplishing an impossible task.

It is interesting to contrast Hallsten's view with Burisch's (Chapter 5). Burisch, in his action theory of burnout, argues that loss of autonomy plays a central role in the process. But Hallsten discounts loss of freedom and instead emphasizes task accomplishment or efficacy. In focusing on self-efficacy as the central factor in burnout, I would tend to favor Hallsten's view over Burisch's. This is not to say that autonomy is not important. Hall (1976) points to auton-omy as a crucial precondition for psychological success, and many research studies have found strong relationships between low autonomy and burnout.

However, autonomy is only important as it relates to control and self-efficacy. Low autonomy usually is associated with burnout because when auton-omy is low there generally are constraints and demands that interfere with goal attainment. Thus, it becomes more difficult to achieve a sense of efficacy. But it is theoretically possible for people to avoid burnout even when autonomy is limited, as long as the limitations on their autonomy do not interfere with the attainment of meaningful goals.

This possibility turns out to be more than just theoretical. In my original research on stress and burnout in novice helping professionals, I studied two groups that differed markedly in their initial autonomy. New public health nurses in my study were put into a kind of apprenticeship position during the first 6 months of their employment. They spent most of their time shadowing a more experienced nurse and were not allowed to work on their own at all for the first month or two. Their autonomy was severely limited. New public school teachers, on the other hand, were "thrown" into a classroom totally on their own. There was little restriction on their day-to-day functioning. Who was more susceptible to burnout?

Of the four professional groups that I studied (nurses, teachers, mental health professionals, and lawyers), the public health nurses experienced less stress and burnout, even though their autonomy was limited, because as novices they needed a great deal of structure and guidance in order to feel competent in their work. Feeling competent ultimately was more critical than feeling free. Of course, as the novices became more experienced, they needed much less direction and structure to feel competent in their work. And it was not too long before restrictions on autonomy began to interfere with achievement of personally meaningful goals. But in the first few months of professional work, self-efficacy and autonomy were at odds. And when they were, it was self-efficacy that became more important in helping professionals to avoid burnout.

Like Hobfoll and Hallsten, Pines (Chapter 3) argues that goal attainment and success are crucial antidotes to burnout. Although she emphasizes "existential significance" rather than self-efficacy in her work, both play a prominent role in her model. In fact, she repeatedly emphasizes that it is impossible to feel that the things we do are meaningful and useful if we fail to achieve important goals. There can be no existential significance without self-efficacy. In her model, burnout follows directly from failure and the sense of helplessness that it breeds.

But Pines, like Hall in his model of psychological success, makes an important point about the relationship between goal attainment and burnout: not all goals are equally important for the prevention of burnout. It is the attainment of personally meaningful goals that alleviates burnout. In other words, one must feel efficacious in areas that are meaningful and significant in order to escape burnout. A careful reading of Bandura's work on self-efficacy suggests that he would concur with this view. He seems to suggest that successful attainment of meaningless goals will do little to increase a person's self-efficacy. However, Pines's and Hall's models are useful in making it more clear and explicit that goal significance is as important as goal achievement in the promotion of self-efficacy and the prevention of burnout.

DOMAINS OF PROFESSIONAL SELF-EFFICACY

Self-efficacy is not a global personality trait. A person can feel efficacious in one kind of role or situation but not another. Thus, in applying the term to the problem of professional burnout, we need to recognize that it is professional self-efficacy that is most important. By professional self-efficacy, I mean professionals' beliefs in their abilities to perform in professional work roles.

Gibson and Dembo (1984) provide an interesting example of how professional self-efficacy can be operationalized. They designed a 30-item "Teacher Efficacy Scale." It included items such as, "When I really try, I can get through to most students," and "If a student in my class becomes disruptive and noisy, I feel assured that I know some techniques to redirect him quickly." Although they did not examine the relationship between teacher efficacy and burnout, they did observe teacher behavior in the classroom and found that high-efficacy teachers spent more time monitoring and checking student work and more time preparing. These teachers also criticized students less and praised them more, and they showed more persistence when students gave the wrong answer.

Teachers, like other professionals, spend much of their time interacting with their "recipients," and this is the focus of Gibson and Dembo's measure of teacher efficacy. These interactions are certainly important potential sources of stress and burnout. However, interactions with others in the work environment—peers, superiors, and a variety of bureaucratic functionaries—can be even more important sources of stress and burnout. Thus, it would seem useful to expand professional self-efficacy to include at least three different domains of professional role performance: task, interpersonal, and organizational.

The Task Domain

The first domain is the one that Gibson and Dembo focus on. It concerns the technical aspects of the professional role. In the case of teachers, it relates to how skillful they feel they are in preparing and delivering lessons, in correcting student performance, and in motivating student effort. In the case of nurses, it would relate to how competent they feel they are in handling intravenous apparatus, administering medications, and so on.

The Interpersonal Domain

The second domain of professional self-efficacy relates to a person's ability to work harmoniously with others, particularly recipients, coworkers, and immediate supervisors. In many helping professions, interpersonal self-efficacy is closely related to task self-efficacy. For instance, the success of a teacher will depend in part on how effective he or she is in developing rapport with students.

The Organizational Domain

The third domain of professional self-efficacy refers to beliefs about one's abilities to influence social and political forces within the organization. For professionals working in complex organizations, this last domain of professional self-efficacy is especially important. Numerous studies have suggested that organizational constraints and demands represent a major source of stress and burnout (Drory & Shamir, 1988; Leiter & Maslach, 1988; Maslach & Jackson, 1984a; O'Driscoll & Schubert, 1988; Pines & Aronson, 1988). If this self-efficacy view of burnout is correct, then professionals' beliefs that they lack the ability to influence these organizational sources of stress must be particularly pernicious.

My own research tends to support this view. The subjects in my study who were most able to overcome early career stress and who were most resistant to burnout during the first decade of their careers seemed to display particularly strong organizational self-efficacy (Cherniss, 1990). They had the belief that they could make an impact on their work settings, and their actions confirmed the belief.

An especially good example of a professional who seemed to have strong organizational self-efficacy was Mark Connor. Mark was a school psychologist who had a very negative initial career experience. He went into school psychology because he wanted to counsel troubled children and adolescents. But he spent almost all of his time administering psychological tests, writing reports, and attending meetings. He also received little support from colleagues or supervisors, and he became embroiled in frustrating conflicts with some teachers and administrators. Mark considered leaving the schools and going into private practice. But he remained a school psychologist, and after 10 more years of practice he was one of the most satisfied professionals in the study.

One reason that Mark was able to overcome early career burnout while remaining in the schools is that he discovered ways of changing the system to allow himself more time to do what he wanted. For instance, at one point a budget shortfall in the district led to the elimination of the elementary school counselors, a decision that was politically unpopular. At that point Mark formulated an innovative solution. He proposed that the school district allow him to establish a counselor internship program. Students from graduate programs in local colleges would work in the elementary schools. They would be supervised by Mark and would receive academic credit. The only cost to the district would be to hire another psychologist part time in order to allow Mark the additional time needed to run the program. Using considerable organizational negotiation skill, Mark shepherded the proposal through all of the appropriate channels, eventually securing the support of the school board.

The internship program benefited the schools, the interns, and many students, but it also proved to be a highly rewarding activity for Mark. His job became more interesting and meaningful. And while Mark was fortunate to work in a system that was supportive of these innovations, change would not have occurred without his belief that he could influence the system. In other words, organizational self-efficacy, the third domain of professional self-efficacy, was critical in Mark's positive career adaptation.

ENHANCING PROFESSIONAL SELF-EFFICACY

If professional self-efficacy is what enables professionals to overcome stress and avoid burnout, then how does it develop? And how might we encourage the development of stronger professional self-efficacy? Again, Bandura's original work on the self-efficacy concept suggests some promising ideas. In fact, to me this is the real virtue of linking burnout with a construct like self-efficacy. Doing so suggests potentially useful new approaches to research and action relating to professional burnout.

Bandura (1982) proposes that the most direct and effective way of enhancing professional self-efficacy is through performance mastery experiences. For instance, professionals who successfully effect change in their organization are likely to believe that they have the ability to do so again in the future.

A second mechanism for developing self-efficacy, according to Bandura, is vicarious experience. To take another example, if a professional sees a colleague, who appears to have similar ability, successfully change the system, then the professional may develop greater professional self-efficacy (see also Chapter 4).

Bandura recognizes that verbal persuasion and other types of social influence represent a third way of enhancing self-efficacy. But he regards this as less effective than either direct or vicarious performance mastery experiences.

These initial thoughts about how to enhance self-efficacy are helpful, but they seem to beg the question of how one ensures that human service professionals experience success in the three domains of professional self-efficacy that I have identified. Pines (Chapter 3) provides a simple but useful framework for thinking about how to promote professional self-efficacy in the work setting: provide resources and remove obstacles in order to help individuals accomplish meaningful goals. Let me now expand on this general guiding principle and suggest some more specific ideas.

Adopting More Realistic Performance Standards

First, success depends in part on internal performance standards. Overly stringent performance standards ensure failure, feelings of inadequacy, and burnout. Early popular works on burnout tended to emphasize this aspect of the problem. Both Freudenberger and Richelson (1980) and Edelwich and Brodsky (1980) admonished professionals not to set unrealistically ambitious goals for themselves. Pines's (Chapter 3) discussion of her research on different groups in Israel tends to support this view.

However, as we have just seen, verbal persuasion has not proved to be a very effective means of behavior change. We need to discover what factors in professional training and work environments influence the process by which professionals develop those internal performance standards that are so important in determining whether they will experience success or failure.

While performance standards are important, and it would seem that self-efficacy could be enhanced by lowering performance standards, the quality of care will suffer if standards are lowered too much. As noted above, goals must be significant as well as attainable. We might make goals more attainable by lowering performance standards, but at some point those goals will cease to be very significant or personally meaningful. Also, it is not advantageous for recipients of service or for society as a whole to have professional standards lowered too much. Thus, once the standards are reasonable, professionals need to have the knowledge and skill necessary to meet those standards. This is where the tripartite model of professional self-efficacy that I presented earlier can be useful, for it points to three distinct domains in which professionals need to be competent.

Changes in Professional Training

All professional training programs attempt to develop competence in the task domain, but there is only so much training that can occur before a professional "becomes" a professional. In my research on new professionals, I found that many felt inadequately prepared for the task domain of their work when they began their careers (Cherniss, 1980a). This sense of inadequacy made them vulnerable to burnout.

Although professional training programs could be improved, an even more effective way to help new professionals develop self-efficacy in the task domain is through more carefully planned orientation experiences. A few of the professionals in my study worked in settings that provided these. One of the public health nurses, for instance, spent the first 2 months on the job "shadowing" a more experienced nurse. During the next 2 months, this novice gradually assumed more responsibility for her own workload.

In another case, a mental health professional began her career working in an alcoholism program. She had little formal training in this area, and so the agency sent her to several training seminars and workshops during her first 6 months on the job. Like the public health nurse, she also spent the first 2 months observing agency staff conducting treatment.

Professional training programs appear to do even worse in preparing new professionals to deal with the interpersonal demands of professional work. And in many professional groups, this lack becomes a major source of burnout. For instance, a recent study of stress and burnout in Israeli teachers found that teacher–student relations were the single most important cause of burnout (Friedman, 1991). The author pointed out that teachers rarely receive training in how to deal with the social aspects of the classroom. He advocated more training programs for teachers in this area.

Stress management and burnout workshops sometimes do provide training relevant to the interpersonal domain. Assertiveness training would be an example. Typically, however, such interventions do little to address self-efficacy in any of the domains. Participants are taught to think differently about interactions with clients, for instance. And they are taught how to manage the symptoms generated by aversive interactions. However, they usually are not taught how to improve those interactions.

It is in the third domain—the organizational—that professional training programs appear to be most inadequate, as Schein (1972) pointed out many years ago. He argued that professionals, no matter what the field, should acquire a better understanding of the social and institutional environments in which they work. Unfortunately, his recommendation has not been embraced by many professional training programs. Given their lack of skill and knowledge concerning how organizations work and how one can work effectively within them, it is not surprising that professionals seem to be particularly low in organizational self-efficacy, and that organizational constraints and demands prove to be such a significant source of burnout.

An interesting direction for future research is to develop programs designed to increase professionals' skills in dealing with organizational problems and to evaluate whether such programs increase professional self-efficacy and reduce burnout. Kramer (1974) already developed an interesting program for training nurses in these skills. I believe that programs like hers could accomplish much in alleviating the problem of professional burnout in the human services. I hope that in the future researchers will look more closely at the impact of such programs on professional burnout.

Career Counseling

Kramer's program, like most stress and burnout interventions, employed a didactic group design. An alternative design worth exploring is career counseling. A basic premise in using this approach is that low self-efficacy and burnout become more likely when there is a poor fit between a person's goals, interests, or skills on the one hand, and his job or organization on the other.

Thus, in career counseling, the individual first is helped to become more aware of his goals, interests, strengths, and weaknesses. Second, the counselor helps the individual to critically examine his career goals and to set more concrete and realistic ones. Third, the counselor helps the individual to identify the general job characteristics that are most likely to enhance self-efficacy and fulfillment. In other words, what does this individual need in a job and work setting in order to feel successful and happy? Finally, the counselor helps the client to develop skills necessary for securing those desirable job and work setting characteristics. For instance, the person may be taught how to be more effective in negotiating with superiors.

Career counseling is more complex and involved than the popular 1- or 2-day "burnout workshop" in which relatively large groups of people come together to participate in didactic and experiential exercises. However, compared to other types of counseling and psychotherapy, career counseling is relatively economical, usually requiring no more than 6–12 one-hour sessions.

Ideological Communities

Thus far I have focused on some individual-based strategies for enhancing professional self-efficacy. But, as noted earlier in this paper, self-efficacy theory recognizes that the environment plays a significant role as well. More specifically, self-efficacy tends to increase in situations characterized by predictability and control (Bandura, 1982). Unfortunately, many professionals work in situations characterized by low predictability and low control.

Up to now, there have been many correlational studies demonstrating a relationship between burnout and role ambiguity or lack of control (Drory & Shamir, 1988; McMullen & Krantz, 1988; O'Driscoll & Schubert, 1988). It would be valuable for researchers in the future to move on to the next logical phase in the research enterprise, namely, field experiments in which ambiguity is reduced while control is increased, and then the effect on burnout is measured. Jackson's (1983) study of nurses provides a good model of such research. Jackson made the work environment of nurses somewhat less ambiguous by training supervisors to hold regular meetings in which important issues were discussed. Her results suggested that the meetings helped to reduce role ambiguity and stress for her subjects.

While Jackson's approach is promising, a few years ago I proposed another, more profound way of reducing uncertainty and conflict in professional work environments (Cherniss & Krantz, 1983). I had studied two professional work settings in which there should have been high levels of burnout, given the kinds of problems with which the staff had to deal. Yet there was remarkably little burnout in these settings. One of the settings was a residential program for mentally retarded people operated by a Catholic religious order. The other was a school for emotionally disturbed children based on the Montessori philosophy. In searching for reasons for the low burnout levels in these two settings, I identified one factor that seemed especially important: both settings were what I came to think of as "ideological communities." By this I mean that there was a very clear, explicit, well-developed, and uniform ideology on which work in the setting was based.

In interviews with the staff in these settings, I became aware of how working in an ideological community can provide one with a framework for thought and action that helps reduce ambiguity and conflict. For example, one of the teachers at the Montessori school told me that before she came to this school she taught in others that did not follow any particular philosophy or model. She had her own values and ideas, but they were vague, general, simplistic, and not based on any external authority. And she felt pressure from other teachers and parents to stress different things in her teaching. Then she came to this Montessori school and almost immediately felt a sense of relief, for the approach gave her the moral support to reject competing demands and expectations once and for all, and to use a consistent approach that made sense to her (Cherniss & Krantz, 1983, p. 200).

This teacher went on to explain how when she adopted Montessori she found herself becoming more relaxed. As she felt more confident and relaxed, her attitudes toward the children became increasingly positive. As this happened, the children seemed to respond. They became happier, more motivated, and easier to teach. They learned more. As they did so, the teacher became even more relaxed and self-confident. The previous vicious cycle of self-doubt, failure, and frustration was replaced by a positive cycle of confidence, competence, and success (Cherniss & Krantz, 1983, p. 201).

While having an ideology or set of beliefs to guide one's work can be helpful in alleviating stress and burnout, this does not seem to be enough. In fact, some ideologies might even exacerbate conflict and stress, as Pines (Chapter 3) discovered in her study of Israeli reactions to the Intifada: members of progressive, left-wing organizations tended to score highest in burnout, even though many of them embraced a strong ideology. In the two settings I studied where burnout was low, there were two additional factors that made the ideologies supportive. First, everyone in the setting adhered to the same guiding ideology. Thus, it was constantly reinforced. Second, in both of these cases the ideology had been translated into specific guidelines and routines. For instance, in the case of the Montessori model, the general philosophy is implemented and reinforced through an elaborate, formal, standardized curriculum that specifies the materials to be used in teaching certain concepts, the arrangement of the classroom, and prescriptions concerning the role of the teacher. Each teaching unit is broken down to small, predetermined steps. In this way, the ideology is made an active, vital force that continually provides support and structure for professional workers in the setting. This approach is similar to the one that Pines (Chapter 3) advocates: goals are made concrete and achievable, people are provided with effective means for achieving those goals, and they have mechanisms that give them almost continuous positive feedback. Organizations that combine a strong, shared ideology with concrete goals, effective work methods, and positive feedback should have much lower levels of burnout because self-efficacy will tend to flourish in such settings. Pines's (Chapter 3) example of the Moked organization seems to be such a setting. Thus, transforming organizational work settings into ideological communities seems to represent a potentially powerful way of reducing uncertainty, enhancing professional self-efficacy, and alleviating burnout.

Social Status

Control and predictability already have received much attention in previous research on burnout. But there is another environmental factor that affects self-efficacy and that has not received much attention to date: social status. Bandura (1982), citing earlier work by Langer (1979), proposed that being cast in a subordinate role or assigned an inferior label has a negative impact on self-efficacy. When I came across this finding, I was struck by the fact that so much of the research and writing on burnout has involved human service professions that tend to be lower in status—nursing, teaching, and social welfare. And in teaching and nursing, there has been a particularly high level of dissatisfaction in recent years with the low status accorded to practitioners in these fields. Thus, another interesting area of inquiry is the role of low status in the development of burnout.

In essence, what I have tried to do in this chapter is to use a single, guiding, theoretical construct to help organize and integrate what we already know about burnout and to provide some specific directions for future research on the problem. I think that in the process of doing this, I also have suggested some promising ways in which burnout might be alleviated.

9

ORGANIZATIONAL STRUCTURE, SOCIAL SUPPORT, AND BURNOUT

Jacques Winnubst
University of Utrecht

Work organizations vary in the degree to which their members get along well or poorly. These variations are probably related to the structure of the organizations, and also to the organizational culture and existing ethics. For an organization to function well, it is very important that each member be willing to do something for another. We call this social support—a concept that signifies the fact that one human being recognizes the other person's identity, values that person, and sometimes actually helps him or her. If an organizational member (employee) lacks adequate social support under conditions of high stress, there is a strong chance that he or she will suffer from strain (Winnubst, Buunk, & Marcelissen, 1988). Stress and strain develop in a sequential process (Marcelissen, 1987), and one of the outcomes is burnout (Gaines & Jermier, 1983).

We see burnout as a work-related problem that can emerge within every occupation (cf. Fischer, 1983; Golembiewski, Munzenrider, & Stevenson, 1986; Meier, 1983). Thus, we have chosen the definition of Pines and Aronson (1988), which states that burnout is a condition of physical, emotional, and mental exhaustion that is the result of chronic emotional strain. This definition does not restrict burnout to individuals who do "people work" (as is the view of Maslach, 1982a). Nor do we follow the older view of Pines, Aronson, and Kafry (1981) that restricted the concept of burnout to health professionals, especially to employees who do people work, and the concept of tedium for all other types of employees. According to Schaufeli (1990), tedium is a condition of extreme fatigue that can be the result of any chronic stressor, but burnout is the result of a process in which emotional strain forms the core. However, in our view extreme fatigue and emotional exhaustion are two different sides of the same coin, which is burnout.

This chapter focuses on the interrelationships between stress, strain, social support, and burnout. However, contrary to most other approaches, we place these interrelationships within the framework of organizational structure and culture. This chapter's contributions to the study of burnout can be summarized as follows: (1) The types of stressors with which burnout are linked covary with organizational structure; blue collar stressors are dominant in the machine bureaucracy, as white collar stressors are in the professional organization. (2) The way social support is provided depends on the organizational climate, which, in turn, is associated with organizational structure. (3) The symptomatology of burnout is identical for blue and white collar employees; however, the etiology of burnout differs according to the organizational type.

The nature and scope of this chapter are more theoretical and conceptual than empirical. Some of the ideas are founded on good empirical research, but others are preliminary and have to be tested. Enlarging the theoretical framework around the concept of social support seems to be necessary. This development will raise new challenges for research.

ORGANIZATIONAL CULTURE, CLIMATE, AND ETHICS

According to Feldman (1988), organizational culture is a collection of meanings that have developed in the context of the organization but that interact with the wider social and cultural context. The members of the organization handle these meanings as norms, roles, plans, ideals, and ideas with the intention of creating coherence and giving sense to the stream of actions and events of which they are part.

How strongly the organizational culture is determined by the structure of the organization may become clear by looking at the well-known Mintzberg (1979) analysis of ideal types. We restrict ourselves here to the two most important and most used types of structure: the machine bureaucracy and the professional bureaucracy (see Table 9-1). The machine bureaucracy is an organization form in which many labor processes have been standardized, in which techno-structure is the key word, and in which only limited horizontal decentralization is possible. A national post office, a steel company, or a giant automobile industry would be examples of a machine bureaucracy. In contrast, in the pro-

TABLE 9-1 Machine Bureaucracy versus Professional Bureaucracy

Machine bureaucracy	Professional bureaucracy
Standardization of work	Standardization of skills
Technostructure	Operating core
Limited horizontal decentralization	Horizontal and vertical decentralization
Much formalization	Little formalization

Source: After Mintzberg, 1979.

fessional bureaucracy, the performers are the key figures, there is both horizontal and vertical decentralization, and only skills have been standardized. Universities, general hospitals, public accounting firms, and craft production systems are the embodiments of the professional bureaucracy.

In the machine bureaucracy the sources of burnout are in the drained, routine labor and in the strong regulation of all work processes. Latitude and growth potential are extremely limited. In the professional organization the sources of burnout are in the lack of structure of the work processes and in the open definitions of territory. In the machine bureaucracy there is the constant risk of underload and apathy; in the professional bureaucracy, overburdening prevails and territory fights are frequent. Organizational structure also has an impact on company culture and ethical climates. In the machine bureaucracy, the dominant cultural traits are perfectionism and conformism. In the professional bureaucracy, the dominant features are creativity and autonomy.

However, the idea of an organizational culture is too broad for our purpose, and we will therefore restrict ourselves to an essential aspect of it: the work climate. The work climate of an organization is the result of what most members see as the typical customs, practices, and procedures. The work climate is often described on the basis of such dimensions as autonomy, latitude, degree of structuring, reward system, warmth, and support. In their review of the relevant literature, Victor and Cullen (1988) made a rough distinction between the following:

1. Structure aspect—with a strong emphasis on the nature of the rules, rewards, and control mechanisms
2. Normative, ethical aspect—with a strong emphasis on such values as warmth and support for one's colleagues and subordinates

It is striking how much attention is being given in the literature to this ethical, moral aspect of organizations. Victor and Cullen (1988) provided a typology of ethical climates, as shown in Figure 9-1. The *ethical criterion axis*

LOCUS OF ANALYSIS

	Individual	Local	Cosmopolitan
Egoism	Self-interest	Company profit	Efficiency
Benevolence	Friendship	Team interest	Social responsibility
Principle	Personal morality	Company rules and procedures	Laws and professional codes

(ETHICAL CRITERION — vertical axis label spanning Egoism, Benevolence, Principle)

FIGURE   9-1 Typology of ethical climates. After Victor & Cullen, 1988.

contains the fundamental criteria that are involved in moral choices: maximizing self-interest, maximizing mutual interests, and sticking to principles. *The locus of analysis axis* contains the societal scale on which the moral choices are made. The chosen role in the organization can be rooted in the individual, local, or cosmopolitan dimension.

Moral choices are quite important to the work climate and for the way in which social support may function. A work climate in which the majority of employees tries to maximize self-interest, within a strong individualistic orientation, has different implications for giving support than a work climate in which the prime objective is to maximize mutual interests for the better functioning of the organization as a whole.

When ethical principles can be rendered as both personal morality and organizational codes and social professional rules, then this will lead to social support that is given from inner conviction, strongly rooted in external ethical codes. When Mintzberg's typology (1979) is compared with that of Victor and

Cullen, 1988), new insights come to the fore: suppositions that look worthwhile enough to be tested. This is a productive way in which to examine the social support and burnout themes in a broader perspective than usual. At this point we are in need of an elaboration of the social support theme; after that we will come back to Mintzberg's views on the one hand, and Victor and Cullen's on the other.

SOCIAL SUPPORT IN RELATION TO STRESS AND BURNOUT

As a result of the work climate, people can find themselves in a downward spiral; they feel lonelier and lonelier, and more and more isolated from colleagues and the outer world. Their social relations become fewer, and depression, burnout, and disease loom on the horizon. The investigation into social support (or rather, into the lack of social support) has developed largely because of the fact that the above-mentioned processes can be studied well with it. On the other hand, there are several indications that too much or too intensive social support may threaten employees' need for autonomy (Winnubst et al., 1988; see also Chapter 4).

Social Support: Definition and Weaknesses

What is meant by social support, and what possible pitfalls are present in using this rather new concept? One of the most widely used definitions of social support is that of Cobb (1976, p. 300), who describes it as "that piece of information which convinces people that others love them and care for them [emotional support], that others respect them and value them [affirmative support], and that they are part of a network of communication and mutual support [network support]." In addition, other forms of support have been distinguished (e.g., by Wilcox & Vernberg, 1985). Informative support is the willingness of other people to state opinions and give information. Instrumental support is the willingness of other people to give material aid (cf. Winnubst et al., 1988).

Payne and Jones (1987) point to a number of weaknesses in the social support concept. They conclude that the following questions are essential for measuring social support:

1. Direction: Is social support given to others, received by others, or are both options the case?
2. Availability: Is social support only a possibility, or is it actually given?
3. Description/evaluation: Is the nature and quality of social support described and/or evaluated?
4. Content: What is the type of social support—emotional, instrumental, informational, and so forth?

5. Network: Who are the sources of support—family, friends, neighbors, colleagues, therapists, and so forth?

The original idea in the social support literature is that social support is undoubtedly beneficial under all circumstances, or at any rate in times of crisis, and that this holds equally true for each type of human being in each social stratum. However, this notion has been seriously challenged. The number of studies in which negative results have been reported is still increasing (Schwarzer & Leppin, 1989; Winnubst, Marcelissen, & Kleber, 1982).

Social support not properly furnished can be menacing and can affect feelings of autonomy. According to the research by Marcelissen (1987), poor social support from one's superior is an important causal factor in the development of stress-related problems among employees at lower organizational levels. At higher organizational levels (i.e., in middle and higher management) this causal link between superior support and strain was missing. Obviously, managers are in a position to solve work-related problems themselves, and support from superiors is irrelevant.

Brickman, Rabinowitz, Karuza, Cohn, and Kidder (1982) observed that even well-intentioned social support could have victimization and stigmatization as side effects. For example, the support could be interpreted as evidence that the employee cannot cope independently. In addition, there are great differences between the abilities of people to mobilize or give support. Such differences are related to personality structure and social skills. Indeed, some persons even seem to be adept at keeping other people at a distance (Winnubst & Van den Bout, 1989).

Social Support and Burnout

There is a growing amount of research in which interaction with others (clients, colleagues, chief or head of department) is a main component in the emergence of strain and burnout. Beehr (1981) hypothesized that many employees are dissatisfied with colleagues because they send important role expectations. He found that individuals who experience high stress blame their coworkers for it. Coworkers were an important source of job dissatisfaction. Pines and Maslach (1978) found that the better the work relations, the more professionals liked their work, the more they felt free to express themselves, the more they were likely to stay healthy, and the more successful they felt at work.

Leiter and Meechan (1986) examined social interactions and burnout using the Maslach Burnout Inventory (Maslach & Jackson, 1986). They found higher emotional exhaustion in professionals who had to concentrate on social contacts within a formally defined work area. Professionals who were ambiguous about their contacts with other staff members and restricted social interactions to colleagues in their own subgroup scored higher on depersonalization.

Cronin-Stubbs and Rooks (1985) examined stress among hospital nurses and found that social support was negatively associated with and predictive of burnout. Davis-Sacks, Jayaratne, and Chess (1985) examined female welfare workers employed in a state department of social services. They found that social support, particularly from supervisors and spouses, was associated with low levels of both burnout and mental health problems resulting from job stress. Constable and Russell (1986) found that among nurses the major determinants of burnout were low job enhancement, work pressure, and lack of supervisor support. Leiter and Maslach (1988) found that nurses mentioned interactions with coworkers 10 times more often than contact with patients as sources of stress. Kahill (1986, 1988) investigated professional burnout among psychologists who had various degrees of experience. She found that social support was related to social support by family and friends, but not to experience within the profession.

All of these studies have concentrated on burnout and social support in professionals. Research on stress, lack of social support, and burnout in samples of blue collar workers is lacking. Partly this is the effect of the narrow definition of burnout (cf. Chapter 4). However, burnout-like symptoms and variables in the blue collar occupations can be found under the label of alienation, tedium, work disablement, absence, and turnover.

ORGANIZATIONAL STRESSORS AND BURNOUT

An important theoretical model used in current research on social support and occupational stress is the so-called Michigan model. This model was developed by members of the Institute for Social Research of the University of Michigan (French & Caplan, 1972; Kahn, Wolfe, Quinn, Snoek, & Rosenthal, 1964) and has been further validated and elaborated in research that has taken place in the Netherlands (De Wolff, 1986; Marcelissen, 1987; Van Dijkhuizen, 1980; Winnubst et al., 1982, 1988).

In this model, stress is seen as a relationship between the individual and the environment. Two kinds of stress may threaten the individual: (1) he or she may feel an imbalance between the demands from the environment and his or her resources to meet these demands; and (2) the environment may not provide sufficient opportunities to fulfill his or her needs. Stressors are those demands in the work environment that are perceived by the individual as being problematic, e.g., workload, role conflict, and future ambiguity. Most of the research using this model has focused on these so-called white collar stressors.

Stressors can lead to strains—all those behavioral, physiological, and psychological processes that occur under the influence of excessive demands and that indicate a disturbance of normal, healthy functioning. Several strains, such as high blood pressure and high cholesterol level, are considered precursors of disease. In job stress research, usually only the effects on minor health aspects (in particular, psychosomatic complaints) are studied. The model further suggests that there are two types of variables that moderate the relationship between stressors and strains: (1) personality and (2) social environment, particularly social support.

Blue and White Collar Stressors

We now have at our disposal most of the ingredients for working out a more general view on stress, organizational structure, and burnout. The only missing link is the difference between blue and white collar stressors (cf. Murphy, 1988).

The demand-control model of Karasek (1979, 1989) provides a sound basis for understanding the sources of blue and white collar stress. According to Karasek, employees who work in high-demand/low-control jobs are likely to report the highest frequency of strain and to suffer the highest level of stress-related illnesses. Low-demand/low-control jobs lead to passivity and boredom. Low-demand/high-control jobs lead to low stress, but a loss of interest in work may be the result. High-demand/high-control jobs lead to acceptable levels of stress and to high participation. Low-control/low-decision latitude is characteristic of blue collar work.

Blue collar work is characterized by heavy work and adverse physical conditions (such as heat, dust, noise, or presence of toxic substances), or by monotonous work done at high speed and without much control and skill required, or by complex information-processing activities that have to be performed under time pressure (Wallace, Levens, & Singer, 1988).

On the other hand, white collar work (professional and managerial) is characterized by other sources of stress (Burke, 1988). The most widely examined factors are the Michigan-type variables: role conflict, role overload, and role ambiguity. In general, many employees who are active in middle and higher management have high decision latitude and high control, and much skill is required to do the job. In terms of the Karasek demand-control model, stress in white collar jobs is lower than stress in blue collar jobs. Burke points to the importance of relationships with others as a major source of occupational stress. Career problems are also important sources of stress, as are work–family conflicts.

Work Stressors and Burnout

In our view, there is no difference in the symptomatology of burnout as it is experienced by workers in various occupations. However, there is a difference between the causes of burnout; essentially, the difference is between blue collar stressors and white collar stressors.

In blue collar occupations, which can be found most frequently within machine bureaucracy, burnout is caused by emotional estrangement, which is an effect of long-lasting low-control/high-monotony types of work. The emotional alienation and demoralization, the physical and emotional exhaustion, and thus the burnout are the effect of a continuous and forced preoccupation with technological/bureaucratic work processes in combination with few personal contacts and low social support. Among professionals, the physical and emotional exhaustion are caused by the continuous struggle with other people. Boundary-crossing problems and domain fights; continuous role problems; lack of support from coworkers, from the supervisor, and from the partner at home; high workload and high responsibility for other persons—all of these create a higher risk of burnout.

SOCIAL SUPPORT, STRUCTURE, AND CULTURE

We will now go back to Mintzberg's (1979) typology of organizations and to Victor and Cullen's (1988) model of work climate in order to make a connection with the social support and burnout literature. We are restricting ourselves to the two best known organizational structures as mentioned by Mintzberg (1979); the machine bureaucracy and the professional bureaucracy.

Roughly speaking, there is a parallel between the Mintzberg distinction and the classic division proposed by Burns and Stalker (1961). These authors placed mechanical and organic regimes opposite each other. In a mechanical regime (a machine bureaucracy), the emphasis lies on hierarchy, on vertical structures of command and cooperation, on the precise definition of rights and duties. In an organic regime (a professional bureaucracy), guidance and driving power emanate much more from mutual values and objectives. Hierarchy and authority play a lesser part, and there is both vertical and horizontal communication.

In a mechanical regime, support systems are determined by superiors. Indeed, the immediate superior is of especially great importance if employees are to function well. There is a strong emphasis on the exact definition of tasks, and utilitarian proportions of exchange are characteristic. Participants are expected to carry out their tasks loyally; however, activities are determined by the instructions and the decisions of the superior. When we apply Victor and Cullen's (1988) model to this organizational structure, then we find ourselves close to (1) the ethical criterion axis of maximizing self-interest and (2) the locus of analysis axis of individual local atmosphere. In the mechanical machine bureaucracy, life is more boring and quiet, and social support is aimed at maintaining predictable rituals to a high degree. These rituals are meant to make the entire organization function adequately and to diminish the risk of burnout.

On the other hand, the climate in the professional organization is characterized by more variety, and responsibility is spread among more people. Here the superior is more of a colleague who helps and advises. At the ethical criterion axis, we find ourselves more at maximizing mutual interests; as far as principles are concerned, more emphasis is placed on the ethos of progress. In terms of the locus of analysis, the moral considerations beyond work are more local and cosmopolitan than individual.

Consequently, there is a major difference in the manner in which social support occurs within these two types of organizations. The mechanical machine bureaucracy provides more ease and security to the employee, but at the same time it causes a greater degree of alienation and human estrangement. If we were to use a term from psychopathology to define the burnout-provoking element in this type of organization, then we would want to speak of an obsessive-compulsive tendency. The repetitive and hierarchical aspects that are characteristic for this type of organizational culture threaten employees' mental health (Kets de Vries & Miller, 1984).

Burns and Stalker (1961) were quick to see the price that is often paid by the employee in the professional organization. Absence of strict rules and regulations can give rise to a neurotic tendency in this type of organization (Lammers, 1983). Openness and lack of structure can lead to a wish for definition of tasks and organizational structure. The mental health–threatening and burnout-provoking element in this type of organization is the floating and open aspect of it.

Our analysis leads us to propose the following hypothesis: instrumental support is associated more with machine bureaucracies, but informational and emotional (affirmative) support is associated more with professional organizations. Four conclusions may be stated:

1. Every organizational structure will have a support system that matches optimally.
2. Every contingent support system will be maintained and amplified by a matching work climate.
3. Structure, support, and culture can be characterized by means of ethical criteria.
4. These ethical criteria will enable us to predict to what extent organizations will cause severe strain and burnout.

TREATING BURNOUT BY SOCIAL SUPPORT INTERVENTIONS

Clinical and organizational psychologists can get insights from this presentation for use in interventions. They are probably already familiar with the contingency theories around the leadership theme (Bryman, 1986; Fiedler & Garcia, 1987). In general, it is a good thing for a leader to master the technique of emotional support, but at the same time its application is not to be desired under all circumstances. A good superior knows that it is better to handle a badly functioning group in a direct, instrumental way. However, the social support theme can be applied in a much wider sense in organizations. We can see that knowledge of the type of organization and of the existing work climate is of critical significance for understanding what social support is and how it works. Such knowledge might even lead to the development of a contingency theory of stress and burnout.

The clinically oriented organization adviser or the organizationally oriented psychotherapist might ask the following questions when social support is considered to be a core mechanism in the treatment of burnout:

1. Are we dealing with an employee from a strong mechanical or a strong organic organizational type?
2. What position does the employee occupy—does he or she belong to the executive staff or not?
3. Can we speak about one-way traffic in social interaction, or can we speak about giving and taking? In other words, can we speak of a justified exchange?
4. How much social support is really given and/or received? In other words, are we dealing with real support or only with supposed support?
5. Is there sufficient room for receiving and giving emotional support, or does the instrumental approach prevail?
6. Is the support system capable of change and improvement?
7. Is there room for those concerned to teach people to receive and give support?

An interesting review on workplace interventions for stress reduction and prevention has been written by Murphy (1988). He distinguishes between union-identified (or blue collar) stressors and management-identified (or white collar) stressors. Improvement of interpersonal relations of blue collar employees is generally in the hands of health agencies. For white collar employees there are employee assistance programs, support groups, and individual therapy. Thus, even in the domain of professional help we see interesting differences in the treatment of blue and white collar employees.

When we distinguish between macro, meso, and micro aspects of social support, then it may be clear that macro aspects are less easy to change and to improve. Social support systems are already fixed in organizational design and are institutionalized to a fixed organizational culture. Less profound changes and adjustments can be tried in the meso and micro sphere by means of group training. As a result of individual counseling and training, it may be possible to influence somewhat the results of structural and cultural influences within organizations. Improving communication in small groups and improving the ability of the individual to give or receive social support are the goals of a large working field, supported by research and literature from both clinical and organizational psychology. Provided these are combined with insights from contingency and organizational typology theories, this is an interesting area of application.

A work environment without burnout is an impossible goal (Pines, 1982a). However, much can be done to reduce the risk of burnout, and much of that risk has to do with organizational structure and organizational culture.

10

IMPACT OF PROFESSIONAL BURNOUT ON CREATIVITY AND INNOVATION

Czeslaw Noworol, Zbigniew Żarczyński, Magdalena Fąfrowicz, and Tadeusz Marek
Jagiellonian University

The phenomenon of burnout has been discussed in the relevant literature since the early 1970s (Freudenberger, 1974). It has been described as comprising long-lasting emotional exhaustion, dehumanization of other people, absence of job accomplishments, lowered job involvement, and chronic physical fatigue. A meaningful conception of burnout is that of Maslach and Jackson (1981a, 1986). According to their definition: "burnout is a syndrome of emotional exhaustion and cynicism that occurs frequently among individuals who do 'people work' of some kind" (Maslach & Jackson, 1981a, p. 1). They argue that "strong emotional feelings are likely to be present in the work setting: it is this sort of chronic emotional stress that is believed to induce burnout" (Maslach & Jackson, 1986, p. 6). In their model, the burnout syndrome has three components: emotional exhaustion, depersonalization, and lack of personal accomplishment.

Numerous studies have ascertained that the burnout syndrome can be identified as a way in which workers react to a work situation (Cherniss, 1980a; Golembiewski, Munzenrider, & Carter, 1983). This reaction depends on both external factors (i.e., the work environment) and internal factors (i.e., individual, personal techniques of coping with stress).

One of the most important external factors is the complex of those that cause organizational stress. Professional burnout appears in the situation where a worker is overloaded by work (which can be the result of poor organization) and where management does not ensure essential freedoms in making decisions directly connected to work tasks (Beehr & Bhagat, 1985; Jackson, 1984; Shirom, 1989). The syndrome can also be caused by poor relationships between coworkers, which usually are a source of misunderstandings and conflicts. All of these factors contribute to a worker's feeling of discontent and work dissatisfaction. In many cases this is the main reason for the professional burnout syndrome.

The potential effects of burnout are serious, both for the individual staff members and for entire organizations. The syndrome plays an important role in absenteeism, turnover, and low morale. Moreover, it seems to be correlated with self-reported indices of personal dysfunction, such as physical exhaustion, increased use of alcohol, and so forth (Maslach & Jackson, 1981a). Paine (1982a) describes the serious effect of burnout as a significant pattern of personal distress that diminishes mental abilities and decreases professional performance. Rossiter (1979) points out that burnout increases rigidity, thus stifling thinking and flexible attitudes. Starlie (1982), in describing the burnout syndrome, argues that "burnout is also resignation." Resignation paralyzes creativity and undermines change for the better.

On the basis of all of this work, it is hypothesized that professional burnout can reduce creativity and innovation. However, a review of the burnout literature does not reveal a clear statement of the relationship between the syndrome and creativity. Maslach (1978b) states that "burnout is not a synonym for a loss of creativity," hinting distinctly that the character of burnout is limited, and that the syndrome is connected only with an occupational area.

The aim of this paper is to provide a theoretical framework for this relationship between burnout and both innovation and creativity. This relationship can be viewed from two perspectives: adaption-innovation theory (Kirton, 1991) and a concept of creative style of behavior (Strzalecki, 1989). Both some theoretical background and empirical studies will be presented, although more attention will be paid to the former than the latter. Our intention is to contribute to the development of burnout theory and to elucidate the role of burnout in the restriction of creativity.

CREATIVITY AND INNOVATION IN ORGANIZATIONS

The terms *creativity* and *innovation* are often used interchangeably in the literature. To avoid misunderstanding, the terms will be used here according to a classic definition. "Innovation is the discovery of a new, novel, or unusual idea or product by the application of logic, experience, or artistry. This would include the recombination of things or ideas already known. Creativity is the origination of a concept in response to a human need—a solution that is both satisfying and innovative" (Beakley & Leach, 1967, p. 402).

In order for an organization to function well, many of its members should from time to time do something more than just that which results from the range of their obligations. If the system limits itself to function only according to the letter of the law, its development will soon cease. Spontaneous, innovative, and creative activities, which often go beyond the rules of one's role and are at the same time necessary to achieve the organization's goals, are the basic kinds of behaviors that should be demanded by the organization (Katz, 1964). It is the great paradox of organizations that they must not only reduce the variability of human behavior to secure an honest performance but at the same time must create possibilities for some variability, and even encourage it. The behaviors demanded by roles must always be supported by some number of activities, the character of which is innovative and rather spontaneous.

The organization that is able to stimulate its members to provide some contribution to its improvement by creating constructive suggestions and ideas will be better than others. It recognizes that employees who are close to operational problems can often give very useful information about those operations. In contrast, the organization that is only based on reproducing behaviors specified by a role is a very weak social system. The system in which there is a lack of a flow of ideas does not make good use of its capital (i.e., its potential). Thus, the survival and efficiency of an organization depends on those vital resources contained within spontaneous cooperation, and innovative and creative behaviors.

Adaptors and Innovators

The organizational processes described above can be viewed in terms of two decision-making styles. According to adaption-innovation theory (Kirton, 1976, 1977, 1980, 1991), each worker can be classified as being more or less adaptive or innovative. Roughly speaking, all individuals can be dichotomized as adaptors or innovators.

Adaptors are characterized as those who "produce a sufficiency of ideas, based closely on but stretching existing agreed definitions of the problem and likely solutions" (Kirton, 1991, p. 209). They are more concerned with doing better than with doing differently. Their effort is directed first of all at making improvements and "doing better." They usually do not undertake any activities in the direction of changing their organizations.

Adaptive solutions utilize intellectual processes that are easier to grasp and that are directly and obviously acceptable to the majority. These solutions contain ideas that can be derived from existing, noncontroversial assumptions, which thus provides favorable support for the status of the adaptors themselves. Even if the solutions fail, their authors (adaptors) are much more likely to be personally acceptable within their organizations. The success of adaptors rests in part on their having established agreement with their bosses and organizational policy. This feeling of security plays an important role in the consciousness of adaptors. Consequently, any failure of proposed solutions does not damage the adaptor because the responsibility for them is shared with colleagues and bosses.

On the other hand, innovators usually change and reconstruct the problems of their work. They prefer less acceptable solutions. Almost all of their energy is spent in finding solutions to problems that are not so usual or common. Their effort is directed toward doing things in new ways, and not toward doing things better or improving them. Innovators' activities usually assume the character of a creative style of behavior. The tasks they performed are heuristic, in contrast to adaptors' tasks which are algorithmic and noncreative (Amabile, 1983). Innovative ideas are usually not closely related to the prevailing paradigms but, to the contrary, are opposed to the consensus of the group. Thus, there is a stark contrast between innovators and adaptors.

Many authors (eg, Mulkay, 1972; Weber, 1970) point out that the aims of bureaucratic organizations are reliability, efficiency, and precision. In such organizations there is a pressure on officials to be conformist, prudent, and disciplined. This creates good work conditions for adaptors. Such an organizational climate contrasts strongly with that which promotes the work of innovators. The latter organizational climate appears more risky, less understood, and less respectful of the views of others, and it introduces changes whose outcomes cannot be easily envisaged.

Creative Style of Behavior

Creative activity can be defined in the simplest way as searching for multiplicity using various modes of searching. There is a specific transformation of information, with particular utilization of rare, peripheral information in the cognitive system (Berlyne, 1965). Creative thinking manifests itself in problem situations that have a divergent character (Guilford, 1967). It consists of two stages: (1) generation of conceptions and (2) evaluation of conceptions.

It seems that the more exploratory one's attitude, the greater the chance for better generation of conceptions. As seen in other research (Marek, Fąfrowicz, & Noworol, 1993), seeking multiplicity that is connected with seeking stimulation correlates highly with the generation of multiplicity with a cognitive and creative aim. One can suppose that a high request for stimulation correlates positively with creativity, mainly in the conceptualization stage. On the other hand, low self-confidence connected with anxiety reduces creativity significantly, probably in the evaluation stage (Barron, 1957; Goudy & Spielberger, 1975; McKinnon, 1964; Pnina & Leiner, 1975).

Psychological problems connected with creativity seemed to gain a new and more clear cognitive basis when presented in the context of research into styles of behavior. Studying those specific forms of behavior, psychologists made no mention of what the individual was perceiving, but focused instead on a formal analysis of certain behavior forms (Bejat, 1975; Messic, 1969, 1973; Nosal, 1979; Shouksmith, 1970). A style of behavior is the manner of realizing activities typical for each individual in external conditions that are changing (Strelau, 1983).

Analyzing the connections between intellectual and individualistic processes, Royce (apparently the first) has called attention to the problem of style structure. He calls it a multidimensional, organized subsystem of processes (relative to cognitive, affective, cognitive-affective, and epistemic styles) with which an individual manifests his or her cognitive and/or emotional phenomena (Royce, 1973). According to Royce, the formulation of Strzalecki proposes a creative behavior model in which one operates with complex units, in a so-called space of psychological features, which are a synthesis of individualistic and cognitive areas (Strzalecki, 1989).

In previous considerations on creativity, one can observe an evident separation of the two main areas having impact on the effectiveness of an individual's performance: the cognitive (intellectual) area and the area of personality (Strzalecki, 1969). Studies were admittedly conducted in such a way as to confirm the impact of those areas on creativity, but the relationships between both of them were minimized. This leads to a false image of human performance caused on the one hand by rational motives and on the other by subjection to the influence of uncontrolled emotions.

Several factors contribute to such a state. Reykowski (1978), writing about the interdependence between emotional and cognitive mechanisms, states that the artificiality of the assumption about the independence of the performance of these two kinds of mechanisms results from the way they are described. It is influenced by the phases of development of the structures in the life of a child. It means that the artificiality is implied by epistemic categories (Facaoaru & Macarie, 1976; Fiske, 1973; Messic, 1973; Royce, 1973). Another significant reason is the excessive atomization of both cognitive and individualistic areas of cognition, which leads to a separation of homogeneous measures, which are (by nature of their measurement) independent of each other, and of measures of personality (Nowakowska, 1970).

Strzalecki's formulation seems to soften the divergence outlined above, introducing a hierarchical model of creative behavior (instead of a morphological one), which is especially useful in collecting information relative to mutually harmonizing cognitive and personal factors. Application of the hierarchical model affords possibilities for thorough study of the links between those factors on different levels of measurement.

According to Strzalecki's concept, the space of psychological features that underlies the creative style of behavior is described by the following 10 dimensions:

1. Strong ego: concentration on problem solving, ability to cope with anxiety due to the situation of the problem, ability to make decisions, low level of neuroticism. This was also studied by Barron (1963a, 1972).
2. Tolerance of cognitive dissonance: tolerance of discrepancy, conflict, incompatibility, and divergence and lack of information). This dimension was considered by other authors too (Barron, 1963b; Berlyne, 1965).
3. Spontaneousness: self-acceptance, life approbation, spontaneity that determinates active programs of life (Beittell, 1964).
4. Flexibility of cognitive structures: originality and independence of thinking, ability to use different types of methods for problem solving. This classical dimension was previously considered by Guilford (1967).
5. Aesthetic attitude: potential ability to find logical, clear, and aesthetic solutions. It has also been discussed by MacKinnon (1964) in the context of self-image.
6. Self-realization: accomplishment tendency, strong motivation for solving distant problems been over long periods, ambition (also Maslow, 1959).
7. Internal locus of control: readiness to be guided according to an internal value system, flexibility of problem analyzing, independence from external pressure. Many authors have discussed this dimension before (e.g., Lefcourt, 1976; Rotter, 1966).
8. Autonomous cognitive motivation: ability to formulate long-term and ambitious goals, satisfaction due to problem solving. A study in this field was also done by Decci (1975).
9. Originality: readiness to generate new and original solutions. This is a classic Guilford (1967) dimension.
10. Nonconformity: nonconventionality, ability to defend one's viewpoint against external pressure, energy of action. This dimension is often measured in the context of studying creativity (e.g., Crutchfield, 1962; Gough & Woodworth, 1960).

These ten dimensions were derived by Strzalecki (1989) on the basis of factor analysis. They are operationalized by 10 subscales. Each subscale consists of several items that describe potentially creative behavior, within the questionnaire "Style of Creative Behavior." This questionnaire covers a broad span of potentially creative personality characteristics and potentially creative behaviors, and is the basis for a general structural model of style of creative behavior.

The model consists of three main psychological domains (see Figure 10-1). These domains are (1) intellectual (flexibility, originality, and fluency of intellectual processes), (2) personality (freedom and originality of personal-

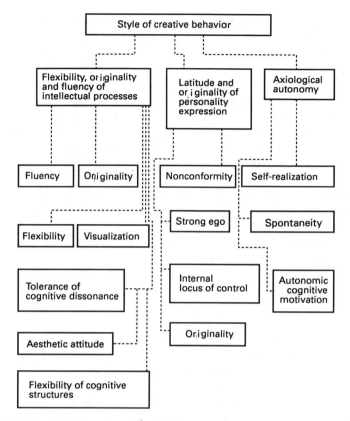

FIGURE 10-1 Strzalecki's (1969) structure of creative behavior style (modified).

ity), and (3) axiological (autonomy of value system). The intellectual domain has been discussed in terms of four classic dimensions: fluency, flexibility, originality of thinking, and visualization (Guilford, 1967). In the current model, it is characterized by three dimensions that correspond with the following three questionnaire subscales: tolerance of cognitive dissonance, flexibility of cognitive structures, and aesthetic attitude. The domain of personality is characterized by four dimensions that correspond with the following four subscales: nonconformity, strong ego, internal locus of control, and originality. The axiological domain is recognized as the domain of the following three dimensions (subscales): autonomous cognitive motivation, self-realization, and spontaneity.

As can be seen in Figure 10-1, flexibility, originality, and fluency of intellectual processes, freedom and originality of personality, and autonomy of value system are characterized by 14 dimensions. Ten of them are operationalized by the 10 subscales of Strzalecki's questionnaire. Flexibility, originality, and fluency of thinking are operationalized by classic tests: the Consequences Anticipation Test (CAT) and the Test of Divergence Thinking (TDT) (Guilford, 1967).

High ratings on the scales of Strzalecki's questionnaire reflect a creative style of behavior. This style seems to be one of the essential factors for the effectiveness and efficiency of innovative behaviors within an organization. If members of an organization are characterized by a relatively high level of creative style of behavior, it can be expected that innovative behavior processes will be effective and efficient. Such workers are called innovators (Kirton, 1991). It is possible to conceptualize the aggregate index of the creative style of behavior for a whole organization. For this purpose all individual rates obtained by each member of the organization on each subscale have to be taken into account. The aggregate index can be very useful for evaluating human resources, in order to estimate the potentiality of creative style of behavior that is accumulated within the organization.

CURRENT RESEARCH

It seems particularly important to study the relationships between the axiological and intellectual domains (i.e., fluency, flexibility, and originality of thinking) of creativity and emotional exhaustion and depersonalization. Moreover, the relationship between reduced personal accomplishment and impoverished self-confidence deserves our attention. Many authors agree that an individual in a state of burnout often has a negative self-image (Gaudinski, 1982). From our point of view this is especially essential because a negative self-image is likely to lead to blocked creativity. This blocking will probably take place on all three traditionally distinguished dimensions of creativity: fluency, flexibility, and originality of thinking.

Subjects and Procedure

These speculations were tested empirically in a sample of 80 male managers, aged 27–46. All managers worked in private or state-owned (chemical, electronic, steel, or construction) companies located in southern Poland. They were employed at the middle-management level and their work experience ranged from 3 to 12 years. Each subject was approached personally and asked to complete the set of questionnaires described below. The scientific purpose of the study was emphasized, and so was the person's anonymity.

Measures

The Maslach Burnout Inventory (MBI; Maslach & Jackson, 1986) was used as a measure of burnout. The three aspects of burnout are measured by separate subscales. High scores on the emotional exhaustion and depersonalization subscales and high scores on the (reversed) personal accomplishment subscale reflect a high degree of burnout.

Several measures of fluency, flexibility, and originality of thinking were utilized in the study. The Consequences Anticipation Test (CAT) and the Test of Divergence Thinking (TDT) are the classic assessments of these basic dimensions of creative thinking (Guilford, 1967). These tests are used quite often in creative thinking research and are characterized by high internal consistency and by correlation with the external expression of creativity.

The main task of the CAT is to invent within 20 minutes as many consequences of 10 uncommon situations as possible. The indicator of fluency of thinking was a sum of all the ideas presented by a subject, and the indicator of thinking flexibility was the number of classes of the ideas presented. The division of the subjects' answers into qualitatively distinct classes was done by competent judges (three psychologists). The indicator of originality of thinking was the number of nonstandard, unusual ideas. Those ideas that were mentioned no more than twice by each 50 subjects were categorized as unusual.

The TDT consists of 16 illustrations similar to geometric figures. The task for a subject was to set the illustrations in order by combining them into classes. A couple of illustrations could be classified as belonging to the same class if there were some particular detail that could be found in each of them. TDT measures fluency, flexibility, and originality of thinking. The indicator of fluency in this test was the number of classes created. The indicator of flexibility was the number of different criteria for classifying the illustrations. Flexibility of thinking is reflected in the skill of developing different criteria for formulating solutions. The division of the subjects' criteria answers into qualitatively distinct classes was done by competent judges (three psychologists). The indicator of originality of thinking in the TDT was the number of classes that were nonstandard but correct. Those answers that satisfied the above conditions and are found very seldom in the investigated population (no more than twice per 50 persons) were an indicator of originality of thinking.

Strzalecki's Style of Creative Behavior Questionnaire (Strzalecki, 1989) was used to assess the remaining dimensions of creativity. Figure 10-1 illustrates the dimensions corresponding with the questionnaire, but the presented structure of creative behavior style has been modified by the authors of this chapter (i.e., only items with factor loadings higher than .35 were included).

The Kirton (1977) Adaption-Innovation Inventory (KA-II) was also administered in order to classify managers as being, more or less, innovators or adaptors. The scale ranges from highly adaptive to highly innovative. According to KA-II, each subject can be located on the continuum based on the level of his or her score.

Results

Cluster analysis using the K means method generated three clusters of managers, based on the pattern of the three MBI variables: emotional exhaustion, reduced personal accomplishment, and depersonalization. These were used as the criterion variables. Four subjects were not included in either of the three clusters. Thus, 76 managers were taken into account in all subsequent analyses. As shown in Table 10-1, managers in cluster I had lower MBI scores than managers in other clusters. In a parallel way, managers in cluster III had higher burnout scores than their colleagues in other clusters. The first cluster consists of managers who are definitely not burned out. Cluster II includes managers who have a low levels of emotional exhaustion and personal accomplishment, and high levels of depersonalization. This cluster includes managers who experience a medium level of burnout, mainly characterized by reduced personal accomplishment and higher depersonalization. Managers experiencing high levels of burnout are those in cluster III.

As can be seen in Table 10-2, a large difference in creative behavior was found between burned-out managers and those who were not burned out. Non-burned-out managers produced three times more consequences for uncommon situations on the CAT than did the medium burnout managers, and almost six times more than the completely burned-out ones ($F = 797.90, p < .001$). The TDT confirms these differences ($F = 55.10, p < .001$) in capacity for invention, although the difference between the three clusters is not as great as the differences on the CAT. Both the CAT and the TDT indicated that fluency of thinking was blocked in clusters II and III. It is worth noting that the significantly lowest values of the tests were obtained in the third cluster. Clearly, professional burnout is strongly related to blocked fluency of thinking, and it may well be the main factor that causes such blocking.

TABLE 10-1   Mean MBI Scores for the Three Clusters

		Mean values of MBI subscales		
Cluster	*N*	EXX	DEP	PA(r)
I	31	12.88	.87	10.22
II	26	13.21	9.56	25.57
III	19	29.43	14.40	24.24

EEX = emotional exhaustion; DEP = depersonalization; PA(r) = personal accomplishment (reversed).

TABLE 10-2 Mean CAT and TDT Scores for the Three Clusters

Cluster	N	Fluency		Flexibility		Originality	
		CAT	TDT	CAT	TDT	CAT	TDT
I	31	64.24	13.25	31.85	8.85	6.17	1.61
II	26	21.00	12.13	17.26	5.12	5.12	.78
III	19	10.36	7.04	8.10	3.46	.76	.34

Similar to the results on fluency of thinking, flexibility differed significantly by burnout cluster ($F = 538.34$, $p < .001$). However, there is an inconsistent pattern of distinction between clusters II and III; the CAT shows a difference and the TDT does not. The first cluster differs significantly from all others; thus, there is an obvious distinction between burned-out and non–burned-out subjects in terms of flexibility of thinking.

Originality of thinking also varied by burnout cluster (CAT: $F = 09.48$, $p < .001$; TDT: $F = 32.25$, $p < .001$). However, this time the difference between clusters II and III was confirmed by both tests. Once again, burnout is strongly related to this dimension of creativity and could be a causal factor in its reduction.

Finally, it was found that almost all managers from cluster I were located at the "innovative" end of the continuum of the KA-II. The subjects from cluster II were spread out along the continuum, but more in the direction of the opposite (adaptive) end. All burned-out managers (with only one exception) placed closer to the highly adaptive end of the continuum.

DISCUSSION AND CONCLUSIONS

These results provide impressive empirical evidence for the significance of burnout in creative behavior. People who are experiencing burnout are characterized by less creativity, on several dimensions, and by an adaptive style of problem solving. In contrast, people who are not experiencing burnout are more creative, on various dimensions, and display an innovative style of problem solving. Considering these data in the context of Kirton's A-I theory, it can be hypothesized that burned-out people will be characterized by the following features: methodical, prudent, conformist, tending to resolve problems rather than find them, dependent, impervious to boredom, accepting of rules, cautious, sensitive, self-doubting, and vulnerable to social pressures and authority.

We presume that the experience of burnout causes the development of these behavioral characteristics; however, other interpretations are possible: e.g., people with these personality traits are more vulnerable to burnout. Further research will be needed to test the viability of these and other interpretations. In addition, the influence of burnout on other features of Kirton's adaptor style (such as precision, reliability, and efficiency) needs further study. Furthermore, long-term burnout may also damage the adaptive abilities of the individual, and not just the innovative ones.

Although our research has established a link between burnout and creative style, the theoretical underpinnings are still unclear. Here we will present some conceptual speculations about how each of the burnout dimensions might be related to the various domains of creative behavior.

To begin, we would predict that reduced personal accomplishment restricts self-realization—the important dimension of the axiological domain of creative behavior. Reduced personal accomplishment is closely related to a negative self-image (Schaufeli and Van Dierendonck, 1992). An individual's negative self-image certainly influences his or her self-acceptance and life approbation. It also invokes a decrease of the spontaneity that determines active programs of life. It can weaken one's ambition and motivation, or even make it impossible to solve distant problems.

Emotional exhaustion and lack of personal accomplishment can weaken one's ability to cope with anxiety and can raise the level of anxiety in the problem situation (Gentry, Foster, Froehling, 1972), as well as the level of neuroticism. Personal accomplishment is related to a tolerance of cognitive dissonance in that lower personal accomplishment causes lower tolerance for discrepancies. Thus, the person will have more difficulties in handling the conflicting demands of different job roles (Lazarus & Folkman, 1984). Furthermore, incompatibility, divergence, and lack of information increase emotional exhaustion, which impairs one's ability to use various methods of problem solving. The readiness to generate new and original solutions is decreased. Emotional exhaustion also damages one's ability to defend one's viewpoint against external pressure. At the same time it minimizes one's energy of action.

Emotional exhaustion may be associated with situations in which there is an excessive inflow of stimulation and insufficient means of regulating it. Many of the human service professions in which burnout occurs are often characterized by just such situations. People working in these job settings are not so flexible in analyzing problems as those who are working with an optimal level of stimulation inflow. Their actions often depend on external pressure, and their readiness to be guided according to an internal value system is decreased. They are reduced to formulating only short-term and unambitious goals. They solve problems that are rather unessential in nature, so that problem solving does not give them satisfaction. Thus, emotional exhaustion blocks two important dimensions of the axiological domain: internal locus of control and autonomous cognitive motivation.

Remaining for long periods of time in conditions of high stimulation initially causes dullness of an individual's sensibility (Grey, 1972). It may also cause the development of depersonalization, as the individual 'shuts out' other persons. This greater depersonalization could block the individual's potential ability to find logical, clear, and aesthetic solutions. In such an individual, a decrease in reactivity will occur (Strelau, 1983) and the need for stimulation will increase (Zuckerman, 1979). After some time, a breakdown in this defensive mechanism will cause a sudden increase of reactivity; possibly this is a cause of emotional exhaustion.

Research has shown that high activity in seeking stimulation correlates positively with the level of creativity (Marek, Fąfrowicz, & Noworol, 1993). A decrease in the request for stimulation, which manifests itself in an increase of reactivity, can block creativity in the conceptualization stage. Burned-out individuals are characterized by a comparatively high level of reactivity that corresponds with low activity in seeking stimulation. Thus, there are some theoretical data pointing to the fact that the burnout syndrome could cause a decrease in creativity within the intellectual and axiological domains.

In conclusion, the professional burnout syndrome can be seen in a new light as an important factor affecting creativity, and innovation in modern organizations. A manager's daily routine is characterized by a need for relevant and fast decision making where creative and innovative thinking is at play (Beakley & Leach, 1967). The well-being of the organization depends on how successfully managers can mobilize their creative and innovative power (as well as the creative and innovative power of their employees). Thus, burnout as a phenomenon that impairs creativity and innovation appears to be an essential factor in determining the effectiveness and development of organizations.

11

BURNOUT, HEALTH, WORK STRESS, AND ORGANIZATIONAL HEALTHINESS

Tom Cox and George Kuk
University of Nottingham

Michael P. Leiter
Acadia University

There appear to be two relatively separate literatures dealing with burnout and work stress. This prompts the question of whether these literatures are redundant in that they simply present different views of the same phenomenon or whether they present different, albeit related, phenomena and can be usefully reconciled. The question is partly a conceptual and theoretical one and partly an empirical one, and its answer requires an exploration of the possible relationships between burnout, work stress, and health. This chapter begins that journey of exploration. It considers the nature of the burnout concept and its relationship with health, and then their relationship to work stress. Finally, it suggests a possible role for organizational healthiness in moderating the processes involved in these relationships. Throughout the chapter reference is made to the argument that different levels of explanation exist for these concepts and that differences here need to be considered in the analysis of the relationships between them.

THE CONCEPT OF BURNOUT

Burnout is used in everyday discourse as a colloquial term describing an emotionally depleted state experienced by people in the helping professions. It first appeared in the scientific literature in 1974, when Freudenberger used the term in an article in the *Journal of Social Issues*. Leiter (1991b) linked the realization of the burnout concept in the United States to the existence of the free clinic movement: "The free clinic movement in the U.S. displayed in an extreme way many of the themes which continue to be central in burnout research. . . . The work environment of the free clinics was characterized by long hours, low pay, emotionally demanding encounters with clients and coworkers, and meager resources. The shortfall in resources was bridged by dedication and comraderie among the staff. . . . The point of Freudenberger's initial paper was that working entirely on the basis of youthful enthusiasm eventually depletes the emotional energies of individuals" (p. 547).

Despite Freudenberger's fathering its introduction to social psychology, it is Maslach and her colleagues (Maslach & Jackson, 1982, 1984, 1986) and Cherniss (1980a), rather than he, who are credited with its development as a scientific concept. Although the concept of burnout is rooted in social psychology (see Chapter 2), it has been imported into clinical psychology, and a tension has come to exist between these two points of view. This has influenced not only the way in which the concept has been cast in different studies but also the way in which burnout has been dealt with in the work situation. The social psychological approach has given rise to interventions focused on the social and organizational work environments while the more clinical approach has predictably focused on the individual worker. To an extent, similar differences in approach are obvious in relation to the management of work stress.

The notion of burnout has caught the imagination of both researchers and practitioners, particularly in relation to human service occupations and the helping professions. Its continuing popularity is witnessed both in the volume of relevant research and in the diversity of its application. By the early 1980s, articles on burnout in refereed journals were being counted in their hundreds (Roberts, 1986) and by the 1990s in their thousands (Kleiber & Enzmann, 1990). Furthermore, Pines and Aronson (1988), in their book *Career Burnout: Causes and Cures*, noted the wide range of different fields to which the concept of burnout has been applied from advanced pathology in schizophrenic patients (McGinness, 1987), through loss of motivation among college students (Meier & Schmeck, 1985), people in the helping occupations (Burke, 1987), and teachers (Kyriacou, 1987).

There are two reasons why the popularity of the concept has grown so rapidly and in the way that it has (Shirom, 1989). First, unlike other possibly related concepts, such as depression or work stress, it does not stigmatize the person, and is thus more conducive to diagnosis and intervention than those that do. The way in which the concept was originally formed, and later empirical evidence, emphasizes its social context and determination: somewhat in contrast, many contemporary theories of work stress focus on the *interaction* between the person and the environment. The person is assigned a more active role in theories of stress than in theories of burnout. As a result, people appear to share more responsibility for their condition (or experience) than in the case of burnout as a social psychological phenomenon. Second and not unrelated, the concept of burnout has provided the social and helping services with what is perceived to be a useful explanation of their own situation during a long period when, somewhat paradoxically, their budgets were being trimmed yet public and political expectations of their role were increasing (see Farber, 1983a; Paine, 1982a). It was useful for them to have a notion that was socially and scientifically acceptable, and that pointed up the effects of this paradox, but without "blaming" those who were affected.

Despite its popularity, however, there are questions that need answering concerning the nature of the burnout concept. On careful examination there would appear to be much conceptual overlap, at least superficially, between the notion of burnout and others, such as work stress and general well-being (health). The question is whether these different concepts are redundant or whether they are different and can be usefully reconciled. Answering this question first requires consideration of the definition and measurement of burnout.

Definition and Measurement of Burnout

The burnout concept was adopted from its lay use to describe a certain group of negative attitudes and emotions that human service workers held toward their jobs (Maslach & Jackson, 1984a). The nature of the concept's origin may have contributed to later problems of definition that have been said to include vagueness and lack of substance, lack of consensus, and overinclusion. These criticisms have also been applied to many other concepts, including work stress. They are often fair when considering the popular literature that tends to the level of everyday discourse, but less so when the focus is narrowed to levels of explanation appropriate to scientific analysis. Notwithstanding this, there are different schools of thought on the nature of burnout and how it should be best measured.

Most of the better known models, such as those based on the Maslach Burnout Inventory (MBI: Maslach, 1982a, 1982c) or on the work of Cherniss (1980) or Leiter (1991a), appear to argue for a concept that variously encompasses three clusters of symptoms:

1. Exhaustion (intellectual, emotional, or physical) and lack of enthusiasm
2. Depersonalization and emotional detachment
3. Reduced personal accomplishment, helplessness, and low self-esteem

Usually clusters of symptoms such as these have been given independent status in multidimensional models (for example, in the MBI: Maslach, 1982a, 1982c, Chapter 2; Leiter, 1991a, Chapter 14). Sometimes they have been more simply grouped together in unidimensional models (e.g., the Tedium Measure: Pines, Aronson, & Kafry, 1981; the phase model of Golembiewski & Munzenrider, 1988). Of course, the scientific status of the burnout concept varies with the type of model adopted.

For multidimensional models, burnout is a "global" term and offers an economy of description in relation to a combination of factors that relate on an additive basis: the whole is the sum of its parts. Here the way in which the term is employed is somewhat analogous to its use in everyday discourse. It is sufficient to imply general meaning but requires deconstruction at a lower level of explanation to allow the operationalization required for scientific study. Such operationalization has been provided through the MBI (Maslach & Jackson, 1981, 1984, 1986), which is the most commonly used multidimensional instrument for measuring burnout. For unidimensional models, burnout represents a single psychological state. Its different aspects are integrated to define a new entity: the whole is greater than the sum of its parts. Here the term doubles for both popular discourse and scientific study.

Questions must arise in relation to the structure of multidimensional models and their core meaning: is one component more representative of the global concept, or the concept as used in everyday discourse, than the others? Furthermore, if there is a key component, we might then ask what are the structural relationships between that component and the others.

Three different types of evidence may be considered in relation to the former question: first, correlations between the scales that represent the multidimensional models and that which represents the unidimensional model; second, correlations between the former scales and assessments of respondents' degree of burnout made independently by others; third, the component profiles of those who declare themselves to be burnt out.

First, there appear to be at least two studies (Corcoran, 1985; Stout & Williams, 1983) that have administered both the (multidimensional) MBI and the (unidimensional) Burnout Measure (BM; initially denoted "Tedium Measure") to their subjects. They report that both the emotional exhaustion and depersonalization scales of the MBI were moderately correlated with the BM (about $r = .50$).

Second, Maslach and Jackson (1981a) compared the scores of police officers on the MBI with the descriptions of the officers offered by their wives. Although emotionally exhausted police officers were described by their wives as coming home upset and anxious, the other two MBI scales appeared unrelated to behavior at home. Rafferty, Lemkau, Purdy, and Rudisill (1986) correlated MBI scores with psychologists' and resident directors' assessments of family practice physicians. Although a correlation between emotional burnout and observed assessments appeared to exist, there was no such relationship between those assessments and depersonalization or personal accomplishment. Ezrahi (1987) compared teachers' MBI scores with structured observations of the teacher's classroom behavior and the assessment of burnout by their superiors. Again, only the emotional exhaustion scale was validated.

Third, Pick and Leiter (1992) reported a relatively small study that administered the MBI to subjects who reported to the authors as a result of a local advertisement asking for volunteers who were burnt out through work. The scores of these subjects on the three MBI subscales were compared with available North American norms (Maslach & Jackson, 1986). Although the subjects' scores on the personal accomplishment and depersonalization scales were not noteworthy in terms of their distributions, their scores on the emotional exhaustion scale were more than two standard deviations above the normative mean. It would appear that these subjects were defining their feelings of being burnt out in terms of extreme emotional exhaustion.

Based on this evidence, and more, both Shirom (1989) and Leiter (1991a) have argued that emotional exhaustion is the defining feature of burnout. Shirom (1989) has argued that future research should focus on this component. The other two components he suggests might be better studied in other contexts. Leiter (1991a), for whom the depersonalization and personal accomplishment components enhance the syndrome's use as a social psychological construct, contends that the three components of burnout should be considered together if their analysis is guided by a theoretical model of their interrelationship.

Leiter (1991a) has described the possible pattern of relationships between emotional exhaustion and the other two components of burnout as defined by the MBI. His study of health care workers in a Canadian mental hospital suggested a strong association between emotional exhaustion and depersonalization, and their relative independence from personal accomplishment. Analysis of his data using LISREL further suggested that work demands such as work overload and interpersonal conflict combined with an underutilization of skill facilitated the development of feelings of emotional exhaustion, which in turn led to depersonalization. Depersonalization was also shown to be influenced by a lack of coworker support, the lack of control coping strategies, and a tendency toward escape coping. Somewhat by contrast, personal accomplishment was determined by skill utilization, coworker support, control coping, and a tendency away from escape coping.

A recent study at Nottingham (Cox, Kuk, & Schur, 1991) considered the factor structure for the MBI in a sample of British nurses and the pattern of relationship between their scores on the scales that emerged from that analysis. The application of principal components analysis to these data suggested a three-factor model of burnout, not dissimilar to that described by Maslach and Jackson (1982, 1986) for the MBI. It accounted for 51% of the data variance.

The most robust of the three factors, in terms of its exact replication across the very different studies, was the emotional exhaustion scale. In the Nottingham study, the nine existing MBI items defined the new emotional exhaustion scale plus one item from the personal accomplishment scale: "I [do not] feel very energetic." The correlation between this new scale and the MBI scale was predictably very strong (Pearson's $r = .99$, $N = 93$, $p < .001$). Nurses' scores on the MBI scales were then intercorrelated. As would be expected, a significant correlation was observed between emotional exhaustion and depersonalization scales (Pearson's $r = .30$, $N = 93$, $p < .002$); no significant correlation was observed between these two scales and the personal accomplishment scale.

Anecdotal evidence suggested that feelings of personal accomplishment or something similar might moderate the relationship between emotional exhaustion and depersonalization. This possibility was formally tested using hierarchical regression analysis (cf. Cox & Ferguson, 1991). The data revealed no such moderating function for personal accomplishment. However, the original suggestion was supported in a subsequent analysis in which a measure of "the meaningfulness of work" (six items: Cronbach's $\alpha = .75$) was shown to moderate the emotional exhaustion–depersonalization relationship ($F = 4.15$, $df = 1$, $N = 93$, $p < .05$). The data suggested that a linear relationship only existed between the two measures when the meaningfulness of work was perceived to be low. There was no such relationship when work was perceived as meaningful.

The evidence reviewed here suggests the centrality of emotional exhaustion in the burnout concept. First, it is what defines burnout to the subject. Second, it is what correlates most strongly with global measures of burnout. Third, it is what determines, in some sort of temporal sequence, at least one other component of the MBI: depersonalization. There is a debate in the burnout literature about the extent to which the burnout concept can be generalized from human service workers to other groups. It is suggested here that it is emotional exhaustion that can be most readily generalized across different work groups, while Leiter (1991b) suggested that it is the depersonalization component which is unique to human service work. If emotional exhaustion is central to the concept, then this is interesting because related concepts are available in work and organizational psychology.

BURNOUT AND HEALTH

Emotional exhaustion readily maps onto the model of suboptimum health (or well-being) developed by the first author during the early 1980s (Cox, 1988c, 1990; Cox, Thirlaway, & Cox, 1984; Cox, Thirlaway, Gotts, & Cox, 1983). Studies at Nottingham attempted to map suboptimum health, as discussed by Rogers (1960), using self-reported symptoms of general malaise. Initially, a compilation of general nonspecific symptoms of ill health was produced from existing health questionnaires and from diagnostic texts used in the United Kingdom. These symptoms included reportable aspects of behavioral, cognitive, emotional, and physiological function, none of which were clinically significant in themselves. From this compilation, a prototype checklist was designed with the experience of each symptom being assessed using a 5-point frequency scale ("never" through "always") that referred to a 6-month response period.

In a series of factor analytical studies on British subjects, now variously reported (Cox et al., 1983, 1984), two factors (or clusters of symptoms) were identified. These factors were derived as orthogonal using varimax procedures. The first factor was defined by symptoms relating to tiredness, emotional lability, and cognitive confusion; it was colloquially termed "worn out." The second factor was defined by symptoms relating to worry and fear, tension and physical signs of anxiety; it was colloquially termed "uptight and tense." This model appeared to have some face validity with British general practitioners and medical and psychological researchers (see Cox et al., 1983). Appropriate scales were then derived (General Well-Being Questionnaire: GWBQ) and have been variously used in a number of studies.

Scores on both scales (worn out, and uptight and tense) have been shown to be determined by individual difference, and by the nature of the person's work and work environment. For example, a study of 300 school teachers revealed that "neuroticism" scores on the Eysenck Personality Inventory were significantly related to scores on the GWBQ. Between 37% and 41% of the variance in well-being was accounted for by neuroticism (emotional instability). However, there was no significant relationship between "extraversion" and well-being (Cox et al., 1983). Significant sex differences have also been reported for workers engaged in semiskilled and unskilled work (Cox et al., 1984). Working women have been shown to report poorer well-being than working men, controlling for the age of the worker. Within this sample, well-being scores were shown to be related to the nature of the work in which the person was employed, e.g., repetitive versus nonrepetitive (Cox, 1990). Scores on the GWBQ have also been shown to influence the person's response to work measured in other ways. Cox, Davis, and Cook (1990, unpublished) showed that the effects of routine computer-based work (data entry) on the report of "muscular aches and pains" are conditioned by the subjects' feelings of being worn out, but not by those of being uptight and tense.

Further development data have been collected by Cox and Gotts through a series of linked studies in Britain and Australia. These have been reanalyzed and the model and its associated scales have been amended to increase their robustness in relation to this international sample (and also to diverse homogeneous samples) (Cox, 1990). A number of symptoms have now been deleted from the two original scales, although no new symptoms have been added. The two new "international" scales are each defined by 12 symptoms but retain their essential nature: worn out, and uptight and tense. The deleted symptoms were among the weaker ones in terms of scale definition (and item loadings). The internal consistency of each scale has been determined and shown to be acceptable (α = .85 for worn out and α = .80 for uptight/tense). Norms have been computed for the international scales based on scale totals.

It is thus suggested that suboptimum health, the grey area between complete well-being and obvious illness, is defined by two dimensions—one related to feeling worn out and the other related to feeling uptight and tense. It has been shown that people vary in the extent to which they report these feelings, both between individuals and across time, and it has been suggested that this variation may not only reflect the nature of the person's work and work environment but may also affect other responses to that work.

WELL-BEING AND BURNOUT

The concepts of being worn out and being burnt out would appear on the face of it to be very similar, but actually they differ in at least two respects. First, they differ in the extent to which the reporting of those feelings is contextualized. The MBI, as a measure of burnout, is effectively socially conditioned, and work provides that social context. The GWBQ, on the other hand, was conceived as a general measure of well-being and thus was deliberately decontextualized in its development. In Warr's terms (1990), the MBI is "job-related" and the GWBQ is "context-free." Logically, the working subject who reports being burnt out should also report being worn out, but the reverse may not be true. General feelings of being worn out may not necessarily be accompanied by burnout. The relationship is asymmetric. Second, both measures are frequency-based, but the MBI is concerned with feelings as they exist at the time of completion while the GWBQ assesses symptom experience during a 6-month period prior to completion. The GWBQ scores would seem to tap the general well-being frame in which current judgments of emotional exhaustion in relation to work are made.

If the relationship between the two measured concepts is logically asymmetric and imperfect, the question of its strength in any particular context must be posed. In the study of U.K. nurses by Cox, Kuk, and Schur (1991), a subsample completed both the MBI and the GWBQ. Their emotional exhaustion scores were correlated with their scores from the GWBQ. A significant correlation was found between emotional exhaustion and feelings of being worn out (Pearson's r = .40, N = 30, p < .02). However, in this context, a correlation of only slightly smaller magnitude was found between emotional exhaustion and feelings of being uptight and tense (Pearson's r = .37, N = 30, p < .03). This pattern of relationship is probably accounted for by a strong correlation between the two GWBQ measures, not always found in other contexts (Pearson's

$r = .52$, $N = 30$, $p < .002$). Partialing out the effects of feelings of being worn out on the relationship between feelings of being uptight and tense and emotional exhaustion significantly reduced that correlation coefficient. The correlations between depersonalization and personal accomplishment, on the one hand, and the GWBQ scores, on the other, were lower still and nonsignificant.

BURNOUT AND STRESS

Both feelings of emotional exhaustion (burnout) and feelings of being worn out (well-being) have been variously linked to the experience of work stress, both as contributing factors to that experience and as possible outcomes (Cox, 1990). The dominant view in burnout research is that a causal chain exists in which the experience of stress contributes to the etiology of burnout, which in turn is related to negative job perceptions, lack of organizational commitment, and withdrawal behavior such as absence and "leaving" (see, for example, Leiter, 1991b). There is good evidence for the second link in this chain (see Lazaro, Shinn, & Robinson, 1984; Leiter, 1991a; Shirom, 1989). Jackson, Schwab, and Schuler (1986) reported a limited longitudinal study of the antecedents and consequences of burnout in 249 teachers using the MBI administered twice by mail across a period of a year. They report that emotional exhaustion predicted, inter alia, their respondents' subsequent thoughts about leaving their jobs and actual leaving behavior. In a somewhat similar design, Shirom (1986) assessed burnout in 404 Israeli teachers focusing on emotional exhaustion and physical fatigue (using an amended version of the BM). Respondents completed questionnaires about a month after the beginning of the school year in 1983 and a month before it ended. Among other things, burnout at the beginning of the year predicted job dissatisfaction and intention to leave teaching at the end of the year.

However, the evidence that stress causes burnout needs to be carefully considered and at two levels: (1) conceptually and then (2) empirically. Conceptually there are three issues. First, the study of any possible relationship between stress and burnout can be carried out empirically, but it is better practice to test out a priori models. Such models require the theoretical exploration of possible relationships before data collection. Second, both stress and burnout can be defined at various levels of explanation. Two were referred to earlier in this chapter: the level of common discourse and a lower (more detailed) level appropriate for scientific definition and measurement. Some studies of the relationship between stress and burnout appear to have operated at the second level in terms of the latter concept, using the MBI to measure its component parts, but have operated at the higher (more general) level in terms of the former. This mixing of levels of explanation must lead to confusion. The third issue relates to the nature of the measures used in relation to that of the concepts being measured. It cannot necessarily be assumed that there is effective equivalence by nature of measure and concept, and that all possible forms of relationship are therefore possible.

Theoretical Integration: Burnout and Stress

At the theoretical level, Leiter (1991a, 1991b) argued that organizational stressors (demands) determine levels of emotional exhaustion, especially when organizational supports, such as social support and skill utilization, are lacking. Such an approach can be easily mapped onto the transactional model of work stress and health developed by Cox (1978, 1985, 1990).

The concept of work stress is one that offers an economy of explanation in relation to the complex perceptual and cognitive processes that underpin peoples' interactions with their work environment and their attempts to cope with the demands of that environment (Cox, 1990). It is a concept that is obvious in everyday discourse, and which at that level attracts a variety of related meanings, but which at the level of scientific analysis takes on a more precise meaning and in doing so deconstructs into a number of component parts.

Work stress has been defined as the psychological state that is or represents an imbalance or mismatch between peoples' perceptions of the demands on them (relevant to work) and their ability to cope with those demands (Cox, 1978, 1985, 1990). This process of appraisal may also take into account the resources and supports available to the person for coping and the constraints placed on coping and on the person's control over the situation. Demands, constraints, and resources may be internal or external to the person. For example, internal demands may reflect the person's needs and expectations, but external demands may arise from the nature of the work that the person does, or the technology with which he or she does it, or the environments in which it is done. Peoples' appraisals of their situation may drive their coping behavior and other more general responses, the success of which feeds back into those appraisal processes.

A work situation that is typically perceived and experienced as stressful is one in which peoples' resources are not well matched to the level of demand placed on them, and where there are constraints on how they can cope and little social support for coping. In addition to any consideration of its situational antecedents or cognitive and perceptual elements, the state of stress is often defined by the person's experience of negative emotion, unpleasantness, and general discomfort, and in the slightly longer term by changes in general well-being. Feeling worn out, and possibly uptight and tense, may result not only from the experience of stress but also from the effects of attempts at coping. At the same time, such feelings feed back and partly determine the experience of, and response to, stress (see Figure 11-1).

The concept of stress represents, at one level, a general notion of distress; at another, a process by which distress occurs; and at a third, the psychological/

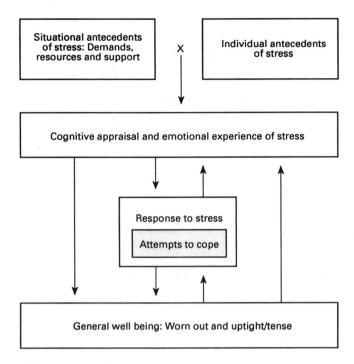

FIGURE **11-1** Transactional model of occupational stress.

experiential state that is central to that process and that itself is a dynamic of that process. The stress state can be described and measured in several different ways, and as a reflection of the person (psychological reasoning) or of the situation (situational reasoning). Thus a number of different possibilities exist: global measures (I feel stressed/the situation is stressful), process measures (measures of demand, control/constraint, support/resources, and abilities, and the interactions between these as internal or external variables), and measures at the level of appraisal outcome (I am anxious or depressed/the situation is anxiety producing or depressing). Finally, the existence of the stress state may be defined for individuals by their experience of negative emotion, and thus this offers an additional type of measure in terms of existing models of mood (Cox & Mackay, 1985; Mackay, Cox, Burrows, & Lazzerini, 1978).

A similar type of analysis, by levels of explanation, can be applied to the burnout concept. It can be deconstructed, in moving from the level of everyday discourse to that of scientific analysis, into its component parts: emotional exhaustion, depersonalization, and lack of personal accomplishment. In turn, these components of burnout may be mapped onto the appropriate level of analysis of work stress. It should be noted that burnout is a condition that is probably unique to human service workers. It can thus only be a particular case within the stress process.

Emotional exhaustion and depersonalization, the two strongly related components that may be central to the concept of burnout, are responses to a particular combination of stressors that affect human service workers. The first is a general response, encompassing elements of emotion and well-being, which appears to give rise to the second, an attempt at coping. The lack of personal accomplishment may have a different status, and be more akin to an appraisal outcome, as might judgments of the "meaningfulness of work'" or Cherniss's (1980a) notion of frustrated professional expectation.

In terms of the transactional model, the burnout concept can be seen to be a particular slice across the stress process that is a relatively common occurrence in human service workers. It is, in the sense of this argument, a mixed bag of an appraisal outcome, an aspect of well-being, and a coping strategy, but one that "hangs together" strongly for this group of workers.

The burnout literature appears more concerned with the situational antecedents of the phenomenon than with its individual determinants. This reflects the origins of the concept in social psychology and the empirical evidence that consistently finds burnout to be more strongly correlated with organizational than with individual variables. Leiter (1991a, 1991b) cites several different types of demand that appear particularly relevant to the etiology of burnout, including issues of work overload, routine nature of work, and interpersonal conflict through work.

Nevertheless, the necessary theoretical overlap between the concepts of stress and burnout suggests that the latter might follow in a particular work context and to some extent from the human service workers' appraisal of their work situation. Such a model is capable of generating specific hypotheses about the relationship between components of the more general stress process and those related to burnout across workers in human services and other occupational groups. To this extent, it is testable. However, most of the studies that have considered the relationship between stress and burnout have not used this type of approach, let alone this particular model. Indeed, most have failed to recognize and deal with the issue of equivalence of levels of explanation. It is not surprising, therefore, that the data available from these studies are equivocal.

Empirical Evidence: Burnout and Stress

What are the available data and, if taken at face value, how might they be interpreted? There are many simple correlational studies relating negative features of work to the experience of burnout. For example, in the Pines et al. (1981) review of research on burnout, there are reports relating burnout to 32 different work-related antecedent conditions, including organizational climate.

However, the directionality of this link has been questioned by the data from (limited) longitudinal designs using cross-lagged correlations. The data from Shirom's (1986) study of Israeli teachers failed to support a causal chain leading from stress to burnout; indeed, for two of his five measures of stress, the reverse appeared true. Similarly, in an earlier study by Wolpin (1988) of 245 school-based educators in Canada, burnout appeared to be a precursor to stress. Furthermore, although significant, the size of the correlations between stress and burnout in many of these studies has been unimpressive (Shirom, 1989).

Why are these data equivocal given the possibility that theories of burnout and work stress can be theoretically reconciled? Several possibilities exist. The first possibility is that there is actually no relationship between the experience of burnout and work stress; the weak correlations are due to unidentified confounding variables. This would seriously challenge the theoretical reconciliation offered above. However, the concepts may be related theoretically but actually unrelated in terms of the measures currently used to tap into them (possibly across different levels of explanation).

Second, burnout and work stress, as generally measured, may be fairly distinct subjective states that arise from similar organizational contexts. Any correlations are due to their shared organizational antecedents. The extent of such sharing may vary across studies. This model is similar to that proposed by Leiter (1991a) regarding the relationship between emotional exhaustion and personal accomplishment. The two dimensions may have little direct impact on one another, but both arise in response to aspects of an impoverished organizational environment.

Third, the relationship between the various components of burnout and work stress processes may be complex and dynamic, as theoretically suggested. Thus, differences in the strength and directionality of the various findings reflect the reality of the situation. The weakness of the correlations may reflect the failure to identify important moderating factors.

STRESS, BURNOUT, AND ORGANIZATIONAL HEALTHINESS

It is unclear as yet which of these possibilities is the more likely, although the second and third sound the more probable. Shirom (1989) suggested that weak correlations between measures of work stress and burnout might be partly due to a lack of consideration and control of possible predisposing and moderating factors (both individual and situational). One source of moderation might be organizational healthiness.

It is generally assumed that the *quality* of the organization, of the work environment, and of work itself can affect the experience of stress and employee health and work performance. That is, the healthiness of the organization may affect the health and performance of its employees. It is the study of this interaction between the healthiness of the organization and health within the organization that Cox termed "organizational health" (Cox, 1988a, 1988b; Cox, Leather, & Cox, 1990). The fact of this interaction offers a chance of promoting occupational health through organizational development—an approach consistent with Leiter's (1991b) comments on dealing with burnout.

Organizational Healthiness

Only a few studies have directly addressed the key concept of organizational healthiness, although there is a somewhat related management literature on organizational effectiveness. Those studies that do exist have largely focused on the development of psychometric instruments for measuring school health (e.g., Childers, 1985; Hoy & Feldman, 1987).

Some of the earliest research was conducted by Miles (1965), who suggested that a healthy organization was one that effectively dealt with its task, maintenance, and growth and development needs. He identified 10 dimensions that reflected the management of these domains. Unfortunately, attempts to operationalize those dimensions have not been particularly successful (see, for example, Kimpson & Sonnabend, 1978).

Hoy and Feldman (1987) suggested an alternative model: "one in which the technical, managerial and institutional levels are in harmony; and the school is meeting both its instrumental and expressive needs as it successfully copes with disruptive external forces and directs its energies towards its mission" (p. 31). School health, they suggested, could be measured in terms of seven different dimensions that together have been shown to be correlated with teacher commitment (Tarter, Hoy, & Kottkamp, 1990).

Cox and Howarth (1990) argued that organizations have both subjective and objective aspects, and that the subjective aspects of healthy organizations have to be in some way consistent with their structure, policies, and procedures (aspects of the objective organization). It has been suggested that the subjective organization can be modeled, by analogy with human physiology, in terms of a number of interrelating subsystems. The healthiness of the organization is a reflection of the "goodness" of its *psychosocial* subsystems and of their coherence and integration.

A recent study of 36 schools and over 500 of their staff in three different U.K. Local Education Authorities (Cox, Farnsworth, Boot, Walton, & Ferguson, 1989; Cox & Kuk, 1991) provided the detail of a model of the healthiness of schools, based on this approach. This study dealt explicitly with the school as a subjective organization and suggested that it could be represented in terms of three psychosocial subsystems. These related to the task environment within the school, the problem-solving environment, and the development environment. Data were collected on teacher stress and absence, intentions to leave, and actual turnover. Measures of burnout were not taken, but teachers' feelings of being worn out were assessed. Teachers were asked to identify their main problems and sources of stress in relation to their work. They were then asked to rate the impact of each of the problems they had cited in terms of their related experience of stress (maximum rating 100). These data were used to calculate a total stress score. There were no differences between types of school or locations in terms of either the number of problems reported by teachers or their total stress scores. Furthermore, there was no relationship between size of school and stress.

Using LISREL techniques (Jöreskog & Sörbom, 1985), Cox and Kuk (1991) considered the relationship between teacher stress and feelings of being worn out, and how this relationship was influenced by the measures of organizational healthiness. The data on teachers' feelings of being worn out are presented in Table 11-1, both as the mean of the scores and a breakdown of those scores by response category equivalents (cf. Cox & Kuk, 1991). These data are consistent with those of Nagi and Davis (1985), and Whiteman, Young, and Fisher (1985), who estimated that close to one-third of their samples of teachers were highly burnt out. The mean levels for being worn out were significantly higher than the available U.K. norms. There were no effects of type of school or its location on well-being.

The LISREL analyses (represented in Figure 11-2) suggest that teacher stress is determined by the goodness of the problem-solving and task environments (with a stronger effect for the former), and that, in turn, teacher stress and the goodness of those two environments determine feelings of being worn out. The strongest effects here are in relation to teacher stress and the goodness of the problem-solving environment. The model that describes this pattern of relationships is associated with acceptable goodness-of-fit indices.

Previous analyses (Cox & Kuk, 1991) had shown that organizational healthiness not only had direct effects on both teacher stress and well-being but moderated the relationship between those two measures. Those analyses, using hierarchical regression techniques (see Cox & Ferguson, 1991), showed evidence of simple (main) effects of teacher stress and of the goodness of the problem-solving and task environments on feelings of being worn out (as reported here). However, there was also evidence of a moderation effect of the goodness of the task and development environments on the relationship between teacher stress and well-being.

These data suggest, among other things, that experiencing a good task environment may attenuate the effects of work stress on teacher well-being. These findings are somewhat similar to those reported earlier concerning the role of the perceived meaningfulness of work in moderating the relationship between emotional exhaustion and depersonalization. Both offer the bases for interventions in work situations to protect employee health. The teacher data were also examined for mediation effects: there were none.

Together these data support the contention that organizational healthiness may both determine the level of work stress and well-being in teachers, and

TABLE 11-1   Distribution of Teacher Well-Being Scores

Range	Worn out
Never to rarely (very low: 0–12)	10.5%
Rarely to sometimes (low: 13–24)	56.2%
Sometimes to often (moderate: 25–36)	30.9%
Often to always (high: 37–48)	2.4%
Means (SD) N	21.6 (7.2) 507
U.K. norms (all)	15.8 (8.5) 1665
Significance of difference:	$p < .01$

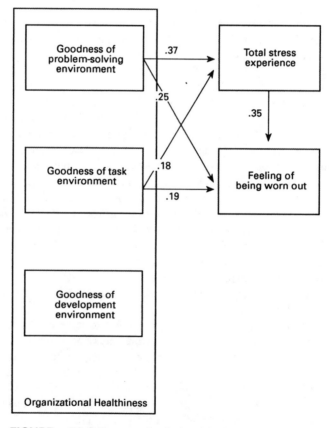

FIGURE   11-2 Structural relationship between organizational healthiness, stress, and feelings of being worn out.

moderate the relationship between work stress and well-being. Given the theoretical arguments advanced earlier in this chapter concerning the relationship between burnout, work stress, and well-being, it is now reasonable to ask whether measures of organizational healthiness can moderate the relationships involved in the burnout and work stress processes.

CONCLUSION

This chapter has attempted to examine the nature of the burnout concept and its relationship to feelings of being worn out, as a measure of general well-being, and to work stress and organizational healthiness. In doing so, it has clarified the theoretical relationship between the burnout and work stress processes, and has suggested a possible role for measures of organizational healthiness in those relationships.

Theoretically, the development of a more detailed model of organizational healthiness in relation to burnout and work stress may help identify the salient characteristics of organizations that have an impact on those processes. Practically, unpacking the role of organizational healthiness in this respect may afford the opportunity to logically plan the promotion of occupational health, and the reduction of burnout and work stress, through organizational development.

It is important to ask whether growth in the concepts of social, work, and organizational psychology represents a refinement of our understanding of the phenomena under study or simply an exercise in redundancy. Early indications are that the latter is not the case for the burnout and work stress concepts. When deconstructed into their different components at equivalent levels of explanation, they can be reconciled. Furthermore, such a reconciliation appears to offer the possibility of usefully extending knowledge in those areas.

IV

METHODOLOGICAL ISSUES

The empirical study of burnout has been hampered by many methodological flaws and fallacies. In fact, problems in this particular area were responsible for the initially slow progress in the study of burnout, during what we have called the pioneer phase (see Chapter 1). Even in the next empirical phase, serious methodological problems were encountered. Recently, however, methodological progress has been made, at least in three main areas: the measurement of burnout, the cross-national study of burnout, and the investigation of the process of burnout.

Schaufeli, Enzmann, and Girault present a state-of-the-art review of the measurement of burnout. It appears that the development of the burnout concept is closely linked with its assessment. This is most convincingly demonstrated by the almost universal acceptance of the Maslach Burnout Inventory (MBI) as *the* instrument to measure burnout. As a result, the test authors' three-dimensional definition of burnout has become the most popular conceptualization of burnout. However, Schaufeli, Enzmann, and Girault show that many more (mostly self-report) instruments exist to assess burnout. As can be expected, there exists a considerable conceptual convergence between these measures despite obvious differences. Most questionnaires focus on mental (and physical) exhaustion—the depletion of emotional resources. In addition to the MBI, special attention is given to the psychometric features of the second most popular instrument: the Burnout Measure (BM). This is a one-dimensional self-report questionnaire that assesses the person's level of (mental, physical, and emotional) exhaustion.

Hence, the choice of the measurement instrument has important consequences for the conceptualization of burnout. It appears from empirical research that the exhaustion dimension—which is generally considered to be the core of the burnout experience—is somewhat unspecific. It is substantively related to other negative mental states.

Golembiewski, Scherb, and Boudreau offer an up-to-date review of cross-cultural work on burnout as well as a critical discussion of the problems that are encountered in this particular kind of research. The authors arrive at nine tentative conclusions. For instance, it seems that the level of burnout is higher among samples from "developing" countries than among samples from "developed" countries. Moreover, indications have been found that different "cultural profiles" are related to particular levels of burnout. In Israeli culture, solidarity and social support are much more valued than in the United States, where competitiveness is a dominant cultural value. From this perspective it is not surprising that Israeli managers are significantly less burned out than their American counterparts.

In the second empirical part of their chapter, the authors present evidence for the cross-national validity of burnout. They demonstrate the cross-national congruence of the MBI dimensions and discuss the distributions of individuals from several national samples across the eight phases of their model of burnout. This "phase model" of burnout hypothesizes that eight progressively virulent phases of burnout exist. Golembiewski, Scherb, and Boudreau also discuss the covariates of these phases in different cross-national settings. Although the results generally agree with those obtained in North American samples (thus supporting the cross-national validity of the phase model), some intriguing differences are found as well. For instance, in the Japanese sample there was a much stronger tendency to depersonalize (i.e., to repress personal feelings) than in the North American or the European, probably because, as the authors suggest, the Japanese seek to "keep face."

In the final contribution, Leiter focuses on the process of burnout. He places burnout in a time perspective and regards it as a developmental process. In doing so, he clearly opposes state conceptions of burnout, which have also been criticized by Burisch and Hallsten in the second section of this book. Leiter starts with the contention that burnout is a three-dimensional construct, including emotional exhaustion, depersonalization, and reduced personal accomplishment. He criticizes the "phase model" of Golembiewski and his colleagues for relying too much on the first dimension and for ignoring the temporal development of burnout. Nevertheless, Leiter acknowledges the contribution of the phase model in classifying individuals, although the underlying dynamic is not considered. In Leiter's own process model of burnout, emotional exhaustion results from a demanding (work) environment, which in turn contributes to increased depersonalization. Accordingly, depersonalization is considered to be a coping response, which occurs *after* the emotional resources have been depleted to a large extent. Reduced personal accomplishment develops parallel with, but separately from, emotional exhaustion and depersonalization.

We hope that the chapters in this final part of the book will stimulate researchers to explore new ground in measuring burnout, in dealing with burnout in cross-national settings, and in investigating process models of burnout.

12

MEASUREMENT OF BURNOUT: A REVIEW

Wilmar B. Schaufeli
University of Nijmegen

Dirk Enzmann
Technical University of Berlin

Nöelle Girault
Réné Descartes University

The importance of reliable and valid instruments to measure burnout is evident not only for the purpose of empirical research but ultimately for individual assessment. Because of the vagueness and overinclusiveness of the burnout concept, researchers were faced with a difficult problem. How is it possible to construct a burnout instrument if the boundaries of that phenomenon are unclear? Soon after the "discovery" of the syndrome in the mid-1970s, dozens of symptoms were associated with it, ranging from *a*nxiety to lack of *z*eal.

In those early years, conceptual confusion prevailed. In the first major review on burnout, Perlman and Hartman (1982) counted more than 48 definitions. They formulated the following "synthetic definition": "Burnout is a response to chronic emotional stress with three components: (a) emotional and/ or physical exhaustion, (b) lowered job productivity, and (c) overdepersonalization" (p. 293). Quite interestingly, their characterization of burnout was rather similar to that of Maslach and Jackson (1981a,b), who a year before had published a burnout inventory that soon became the most widely used instrument. Probably, the recurrently cited summary of Perlman and Hartman with its appealing definition encouraged the use of the Maslach Burnout Inventory (MBI) by providing a conceptual justification. In any case, most burnout researchers were no longer trapped in the vicious circle of an ill-defined construct that could not be measured because it was not properly described.

Use of the MBI implies acceptance of the definition provided by the test authors. Maslach and Jackson (1981a, p. 99) describe burnout as a three-dimensional syndrome characterized by emotional exhaustion, depersonalization ("negative, cynical attitudes and feelings about one's client"), and reduced personal accomplishment ("the tendency to evaluate oneself negatively, particularly with regard to one's work with clients"). In their view the burnout syndrome is restricted to individuals who do "people-work" of some kind. Thus, as a consequence of accepting this definition, the controversy about the nature of burnout was settled by a silent agreement among researchers, at least as far as the human services professions were concerned. This development illustrates how closely the measurement of burnout is related to conceptual issues that have been discussed in Chapter 1 of this book.

The purpose of this chapter is to present a comprehensive overview of measures that have been employed to assess burnout. Several psychometric criteria are used for assessing the various measures, such as inter-item reliability, test–retest consistency, and complexity of the factor structure. Moreover, two types of validity information are considered: (1) consistency with self-report indicators and (2) all too rarely, consistency with indicators not based on self-report. Because most psychometric development has occurred with the MBI and with the Burnout Measure (BM) (Pines, Aronson, & Kafry, 1981), these instruments will be discussed in greater detail.

ALTERNATIVE MEASURES OF BURNOUT

Since the emergence of burnout almost two decades ago, several different kinds of instruments have been proposed to measure burnout. Originally, the syndrome was described rather unsystematically by what can be called clinical *observation* (see also Chapter 1). For example, one of the first people to use the term "burnout" was Freudenberger, a psychoanalytically oriented psychiatrist who observed many mental and physical symptoms among the free clinic staff members with whom he was working. Despite this initial clinical base, a systematic observation method to assess burnout has never been developed. However, a structured *interview* to assess the level of burnout was developed by

Forney, Wallace-Schutzman, and Wiggers (1982). Unfortunately, this approach has not been followed by other researchers.

Another idiosyncratic method to assess burnout involved *projective drawings* (Haack & Jones, 1983). Twenty-six nurses were asked to draw how burned out they felt, and each drawing was independently rated by two psychologists on a 4-point rating scale, ranging from "not burned out" to "very burned out." The nurses were divided into a high-burnout group and a low-burnout group, based on their scores on the Staff Burnout Scale for Health Professionals (SBS-HP; see next section). As expected, the high-burnout group drew pictures of burnout that were rated as expressing significantly more burnout than the low-burnout group. Common themes in the high burnout pictures were exhaustion, isolation, regression, powerlessness, being broken or injured, and feeling overwhelmed. Although empirical testing yielded positive results, this method is not very promising because the criteria for burned-out drawings are unclear. As a result, the ratings may be quite unreliable. Perhaps this is why the authors do not mention inter-rater reliabilities.

An interesting validity study employed an overall *self-assessment* of burnout (Rafferty, Lemkau, Purdy, & Rudisill, 1986). Seventy-six family practice residents were requested to describe themselves over the past several months, keeping in mind the following definition of burnout:

> *The tendency for committed physicians to lose enthusiasm for their work and to become less effective in managing the stress of emotional contact with patients. Symptoms may include some of the following—fatigue, withdrawal from patients and colleagues, cynicism, irritability, difficulty relaxing off work, physical manifestations of anxiety and depression, and feelings of diminished enthusiasm and effectiveness at work. (p. 489)*

Responses were in the form of a 9-point rating scale that ranged from "not at all burned out" to "very burned out." Moreover, each physician completed the MBI. The physician's self-report of overall burnout was correlated moderately with MBI emotional exhaustion ($r = .48$ and $r = .59$ for the frequency and intensity dimensions, respectively) and slightly with the intensity dimension of the depersonalization scale ($r = .34$). These results suggest that physicians assess their global burnout particularly in terms of emotional exhaustion.

In a carefully designed multitrait-multimethod study, Meier (1984) investigated the construct validity of burnout in a sample of 320 male and female faculty members at a university. He included self-ratings of burnout, along with other measures. Subjects were presented with a brief description of burnout and were asked to indicate on a 7-point rating scale the extent to which the construct assessed their current state. This single-item burnout self-rating was moderately correlated with the MBI sum score ($r = .65$) and with two other scales (see next section): the Meier Burnout Assessment scale ($r = .63$) and the Emener–Luck Burnout Scale ($r = .66$).

Most of the above-mentioned methods for measuring burnout are atypical and have only occasionally been used by individual researchers. The rating scales are an exception, however. Such single-item ratings have been employed more or less successfully as a criterion measure to validate the multi-item inventories that have been the most popular way to assess burnout. Therefore, the remainder of this chapter is devoted to this type of measurement instrument.

BURNOUT MEASURES WITH LIMITED APPLICATION

Self-report burnout measures can be distinguished according to the amount of psychometric research on which they are based. For instance, do-it-yourself inventories have not been investigated empirically at all. Most other self-report measures have been studied occasionally, whereas two instruments—the BM and the MBI—have been examined most extensively. Accordingly, these particular instruments will be discussed separately in the next two sections.

Do-it-yourself inventories of burnout have been published under such appealing headings as: "How burned-out are you?" (Bramhall & Ezell, 1981); "What's your burnout score?" (Steward & Meszaros, 1981); "The burnout-test—Examine your beliefs about work, about leisure, about yourself" (Daily, 1985); "The National Job Burnout Survey" (Veninga & Spradley, 1981). None of these inventories has been studied empirically in any way. Therefore, their use is rather limited. At best, they convey a clearer picture of the author's definition of burnout. At worst, the subject becomes alarmed unduly. The best known do-it-yourself burnout scale was proposed by Freudenberger and Richelson (1980, pp. 17–19). They formulated 15 questions, including: "Do you feel fatigued rather than energetic?", "Are you increasingly cynical and disenchanted?", and "Does sex seem like more trouble than it's worth?" The authors also provide an interpretation of the scale scores: individuals scoring in the lowest category are considered to be "doing fine," whereas others are informed that they are "a candidate for burnout" or that they are in fact "burning out." Individuals scoring in the highest category are warned: "You are in a dangerous place, threatening to your physical and mental well-being." However, the "norms" and the corresponding interpretations are not empirically based and should therefore not be taken as valid indicators of an individual's level of burnout.

A number of other burnout inventories have been employed exclusively in a single study. An example of a purely inductive approach is offered by Blostein, Eldridge, Kilty, and Richardson (1985). They factor-analyzed a set of 49 "burnout indicators" in a sample of 400 child welfare workers and found six underlying dimensions: (1) "classic" burnout (e.g., cynicism, depression); (2) negative feelings about clients (e.g., intolerance, postponement of client contacts); (3) a feeling of being overstimulated (e.g., inability to relax, problems with sleeping); (4) a feeling of being overwhelmed (e.g., need to be alone, feeling pulled in all directions); (5) physical problems (e.g., colds and flu, headaches); and (6) lack of intimacy (e.g., emptiness, feeling unappreciated). Typically, in this type of exploratory study, burnout has been identified as a multidimensional concept. However, the obtained dimensions have not been specified theoretically.

Other studies have yielded more interesting results, particularly when additional burnout instruments have been included or when the burnout instrument has been designed for specific occupational groups. For instance, Farber (1984) developed a 64-item *Teacher Attitude Scale* (TAS), which is a modified version of the MBI. The 25 items of the original version of the MBI were augmented with 40 additional items of exclusive relevance to teachers. A factor analysis in a sample of 365 suburban teachers yielded three factors: (1) general feelings of burnout (mainly emotional exhaustion), (2) commitment to the teaching profession, and (3) gratification in working closely with students.

The *Staff Burnout Scale for Health Professionals* (SBS-HP) is a one-dimensional 30-item measure (Jones, 1980b). Twenty items assess burnout, and the remaining 10 items form a lie scale to detect tendencies to "fake good." According to the test manual, the SBS-HP assesses adverse cognitive, affective, behavioral, and psychophysiological reactions that are considered to constitute the burnout syndrome. The SBS-HP (α = .93) has been validated in a series of studies with health professionals (Jones, 1980a). SBS-HP scores were found to be related to external stressors (e.g., high patient-to-staff ratios, high-trauma jobs, working undesirable shifts, and receiving little family support) and to stress reactions (e.g., job turnover, absenteeism, illness, job dissatisfaction, and tardiness rates). However, the samples of the validity studies are very small, ranging from only 27 to 36 health professionals. Moreover, the sizes of the correlation coefficients differ substantially across these studies. More interesting results have come from a larger study of 135 female human service professionals (Brookings, Bolton, Brown, & McEvoy, 1985). Moderate correlations were found between the SBS-HP score (α = .88) and MBI emotional exhaustion (r = .65) and MBI depersonalization (r = .54), whereas the relationship with personal accomplishment was less strong (r = $-.33$). Furthermore, SBS-HP scores were positively correlated with job-related tension and negatively correlated with job satisfaction, internal locus of control, and self-esteem ($.37 < r < .61$).

In a previously mentioned validity study, Meier (1984) employed a self-constructed 23-item true–false burnout test: the *Meier Burnout Assessment* (MBA). Unfortunately, the author does not provide any information about the content of the items. The MBA (α = .79) correlated moderately with the MBI sum score in a sample of 320 male and female faculty members (r = .61). In another sample of 120 students, a similar correlation between the two measures (r = .58) was found (Meier & Schmeck, 1985).

Ford, Murphy, and Edwards (1983) constructed a 15-item burnout inventory that was meant to be used outside the human service professions. The items of their *Perceptual Job Burnout Inventory* (PJBI) reflect (1) emotional exhaustion and cynicism; (2) demoralized, frustrated feelings and reduced efficiency; and (3) excessive demands on energy, strength, and resources. Separate factor analyses were done for a social services sample (N = 237) and a corporate sample (N = 150). In the former sample five factors emerged ("winless," "supportless others," "controlless," "supportless organization," and fatigue), whereas in the latter sample only two factors were found ("supportless organization" and emotional exhaustion). Cronbach's α values for the scales ranged from .67 to .86 in both samples. Obviously, the PJBI cannot be used across different occupational groups.

The *Emener–Luck Burnout Scale* (ELBOS) consists of 30 items and has been employed in a sample of 251 professional human service providers (Emener, Luck, & Gohs, 1982). The internal consistency of the scale is satisfactory (α = .88). A principal components factor analysis yielded seven factors, of which four were significantly related to the burnout self-rating that was used as a criterion measure. These factors are work-related feelings, work environment provisions, dissonance between the individual's perception of self and others' perception of self, and job alternatives. Neither the reliability coefficients of the scales nor their intercorrelations are presented.

Garden (1987a) employed a four-item *Energy Depletion Index* (EDI) in a sample of 95 MBA students with significant work experience. She argued that energy depletion is the dimension of the burnout experience for which there is most definitorial agreement. The EDI (α = .82) is weakly correlated with two aspects of depersonalization: distancing and hostility (r = .25 in both cases).

In a similar vein, Shirom and Oliver (1986) define burnout as comprising physical, cognitive, and emotional exhaustion, and the wearing out of resources. In a cross-lagged panel study among 404 Israeli teachers, they employed a six-item *Burnout Index* (α = .86; stability across 7 months r_t = .70) that correlated .45 with self-reported somatic complaints and .35 with intrinsic job dissatisfaction.

What is common to all the above-mentioned measures is that they assess feelings and emotions that are generated in work-related settings. This applies also for the BM and the MBI. In other words, burnout is considered exclusively on an individual level by the vast majority of the self-report inventories. However, some instruments also assess the situation on the job, i.e., the individual's perception of the organization and its employees. Such instruments that assess both individual and organizational aspects of burnout have been developed for nurses (*Nursing Stress Scale*—NSS; Gray-Toft & Anderson, 1981), teachers

(*Teacher Stress Inventory*—TSI; Fimian, 1984), psychologists (*Psychologist's Burnout Inventory*—PBI; Ackerley, Burnell, Holder, & Kurdel, 1988), and medical personnel (*Medical Personnel Stress Survey*—MPSS; Hammer, Jones, Lyons, Sixmith, & Afficiando, 1985).

Some promising findings about the convergent validity of the TSI are provided by Fimian and Blanton (1987). There was a moderate positive correlation ($r = .64$) between the MBI and TSI total scores in a sample of 413 teacher trainees and first-year teachers. Hammer et al. (1985) report a similar relationship ($r = .69$) between the MPSS and the SBS-HP in a sample of 116 employees from hospital emergency rooms. It is most likely that the exhaustion subscales that are included in all burnout measures are responsible for this positive relationship. Unfortunately, in both studies no correlations between the subscales of the various burnout measures were presented. However, such information is provided by Ackerley et al. (1988), who found in a sample of over 500 licensed psychologists that the PBI overinvolvement scale was most clearly related to MBI emotional exhaustion ($r = .45$), whereas PBI lack of control was negatively correlated with MBI personal accomplishment ($r = -.55$). The remaining two PBI scales (negative clientele and lack of support) were only weakly correlated with the MBI dimensions.

THE BURNOUT MEASURE (BM)

The second most widely employed burnout questionnaire after the MBI is the Burnout Measure (BM) (Pines & Aronson, 1988). According to the test authors, burnout is defined as "a state of physical, emotional and mental exhaustion caused by long-term involvement in situations that are emotionally demanding" (p. 9). In this current view, burnout is not restricted to certain professional groups. Originally, however, the authors had distinguished between burnout and tedium, which were considered to be similar in symptomatology but different in origin: "*Tedium* can be the result of any prolonged *chronic pressure* (mental, physical, or emotional); *burnout* is the result of repeated *emotional pressure* associated with an intense involvement with people over long periods of time" (Pines, Aronson, & Kafry, 1981, p. 15). In their recent work, Pines and Aronson (1988) abandoned this differentiation by broadening the burnout concept, which now also includes tedium. As a consequence, the former Tedium Measure is currently denoted Burnout Measure.

The BM consists of 21 items that express exhaustion and that are scored on a 7-point rating scale, ranging from "never" to "always." The authors present the BM as an instrument for the self-diagnosis of burnout and offer interpretations for test scores. Occasionally, the proposed norms have been used to classify people who are presumably at risk for developing burnout (Aström, Nilsson, Norberg, Sandman, & Winblad, 1991). However, such use for individual assessment is not recommended because valid norms of the BM are lacking.

Nevertheless, the BM seems to be a useful research instrument, as is demonstrated by the psychometric findings obtained by the test authors in 30 different samples totaling 3,900 subjects (Pines et al., 1981, pp. 202–222). Seven years later, Pines and Aronson (1988, pp. 220–222) state that meanwhile more than 5,000 subjects with various backgrounds have completed the BM (e.g., human services, business and management, science, art, administration, technical, clerical, teaching, students). However, most of these subjects come from nonrandom samples that included persons who voluntarily participated in burnout workshops. Accordingly, the possibility cannot be ruled out that selection effects influenced the psychometric results reported by the test authors.

The BM appears to be a reliable instrument with internal consistency coefficients α ranging from .91 to .93. This is not very surprising because many items can be considered synonyms (e.g., "feeling 'burned out'" and "feeling rundown") or antonyms (e.g., "being happy" and "being unhappy"). Thus, for reasons of economy, the length of the BM can be shortened by half of its items without negatively affecting its reliability (Schaufeli & Peeters, 1990). Test–retest reliabilities of the BM range from .89 to .66 across a 1- and 4-month interval, respectively (Pines & Aronson, 1988, p. 220).

BM scores have been found to be negatively related with satisfaction from work, from life, and from oneself, in 30 samples (Pines et al., 1981, p. 209). The mean correlations across these samples are $-.35$ (range: $-.31$ to $-.63$), $-.40$ (range: $-.37$ to $-.70$), and $-.50$ (range: $-.34$ to $-.73$), respectively. In a sample of 129 social workers, BM scores were correlated with an intention to leave the job ($r = .58$) (Pines & Aronson, 1988, p. 221). Also, it was shown in a study of 14 residential facilities for the developmentally disabled that mean turnover rates were higher in facilities with high BM scores than those with low BM scores (i.e., 49% against 17%) (Weinberg, Edwards, & Garove, 1983). In studies among 181 telephone operators and 298 police officers, BM scores were found to be positively correlated with poor self-reported physical health ($r = .48$) and with on-duty physical symptoms such as headaches, loss of appetite, nervousness, backaches, and stomach aches (r values ranged from .32 to .38) (Pines & Aronson, 1988, p. 221). Finally, burnout as measured by the BM is positively related to several work features in a number of samples (e.g., lack of social support, lack of autonomy, lack of feedback, and lack of variety). Typically, correlations with work features are modest, seldom exceeding values of .40 (Pines et al., 1981, pp. 213–218).

Despite the multidimensional definition of the burnout syndrome by the test authors, the BM is conceived as a one-dimensional questionnaire. Thus one can argue that the BM is not a proper operationalization of their definition of burnout. Therefore, it is not surprising that two factorial validity studies have failed to distinguish more than one burnout dimension in the BM (Corcoran, 1986;

Justice, Gold, & Klein, 1981). On the other hand, Enzmann and Kleiber (1989) found some weak indications for a three-factor structure of the BM in a sample of 130 German human services professionals. However, their three-factor structure (i.e., "demoralization," "exhaustion," and "loss of motive") did not correspond to the original components distinguished by the test authors. Recently, Schaufeli and Van Dierendonck (1992) replicated this finding using confirmatory factor analysis in a sample of 667 Dutch nurses.

Unfortunately, only a few studies exist in which other burnout measures, such as the MBI, have been employed simultaneously with the BM (Corcoran, 1986; Stout & Williams, 1983). The results of both of these studies (which have been conducted with 139 MA level social workers and 78 mental health workers, respectively) are quite comparable, however. Burnout as measured with the BM is strongly associated with MBI emotional exhaustion and MBI depersonalization ($.50 < r < .70$), and is somewhat less strongly but negatively associated with MBI personal accomplishment ($-.25 < r < -.30$).

It can be concluded that the BM is a reliable and valid research instrument that indicates the individual's level of exhaustion, which is considered to be the core element of the burnout syndrome (Shirom, 1989; see also Chapter 11).

THE MASLACH BURNOUT INVENTORY

As we have seen before, the MBI test authors Maslach and Jackson (1981a, 1986) described burnout as a three-dimensional syndrome characterized by emotional exhaustion, depersonalization, and reduced personal accomplishment. Moreover, in their view the burnout syndrome is restricted to professionals who work with recipients in some capacity. Accordingly, the MBI is meant for use in this particular occupational context.

Problems can arise when the MBI is employed outside the human services professions. In that case some items must be reworded. For example, Iwanicki and Schwab (1981) demonstrated in a sample of 469 teachers that the MBI can be slightly adapted without any problems for use among teachers by substituting "students" for "recipients." In that case, the meaning of the scales remains unchanged. In a special supplement to the test manual, the "MBI Form Ed" for measuring educator burnout is discussed by Richard Schwab (Maslach & Jackson, 1986, pp. 18–22). On the other hand, Garden (1987a) argued that substituting "coworkers" for "recipients" (as was done by Golembiewski, Munzenrider, & Stevenson, 1986) changes the meaning of the items. As a consequence, the depersonalization and personal accomplishment scales, in particular, have to be interpreted differently.

It is important to note that the three dimensions of burnout have not been deduced theoretically before the proper test construction of the MBI commenced. Instead, they were labeled *after* a factor analysis of an initial set of 47 items in a heterogeneous human services sample (see also Chapter 2). Accordingly, an inductive, rather than a deductive, approach was employed. Burisch (1984a) argued that the latter strategy renders less optimal psychometric results than the former strategy. Thus, despite the almost universal acceptance of the MBI, a careful psychometric evaluation is by no means superfluous, particularly when the inventory is applied in other occupational or cultural settings. To date, elaborate psychometric analyses of the MBI performed outside the United States are completely lacking. (See Chapter 13) Originally, the MBI consisted of 25 items, distributed across four scales: emotional exhaustion (9 items), depersonalization (5 items), personal accomplishment (8 items), and involvement (3 items) (Maslach & Jackson, 1981b). However, the latter scale was presented as an optional part of the MBI. In the second version of the test manual the involvement scale is no longer included (Maslach & Jackson, 1986).

Also, in the second version of the test manual the original double rating of each item has been abandoned. Initially, every item was scored twice on a frequency scale, ranging from "a few times a year" (1) to "every day" (6), and on an intensity scale, ranging from "very mild, barely noticeable" (1) to "very strong, major" (7). Because the frequency and intensity ratings appeared to be highly correlated ($r > .80$), only the frequency scoring is recommended in the latest version of the test manual.

The *internal consistency* of the three MBI scales is satisfactory, with Cronbach's α values ranging from .71 to .90 in the normative sample that includes over 11,000 subjects (Maslach & Jackson, 1986). Generally, similar α values have been found in other samples such as psychologists (Ackerley et al., 1988; Huberty & Huebner, 1988); human services professionals (Brookings et al., 1985); teachers (Belcastro, Gold, & Hays, 1983); social workers (Corcoran, 1986); nurses (Constable & Russell, 1986); and prison guards (Dignam & West, 1988; Lindquist & Whitehead, 1986). Occasionally, internal consistency coefficients lower than .70 are found, particularly with the depersonalization scale in non-human services samples, such as gifted students (Fimian, Fastenau, Tashner, & Cross, 1989) and university students (Gold, Bachelor, & Michael, 1989; Powers & Gose, 1986). As mentioned earlier, depersonalization items may have a different meaning in such samples. However, because poor internal consistency coefficients of the depersonalization scale are sometimes also found in human services samples (e.g., Leiter & Maslach, 1988), the scale's shorter length may also be a factor (because the value of the α coefficient depends on the number of scale items). A standardization for test length eliminates these systematic differences in reliabilities between the depersonalization scale and both other MBI dimensions (Schaufeli & Peeters, 1990).

Test–retest coefficients range from .60 to .80 across short periods up to a month (Maslach & Jackson, 1986). Dignam and West (1988) used a structural equation model in a sample of 200 prison guards and found a "true" autocorrelation of .80 of the composite emotional exhaustion and depersonalization score across a 3-month interval. Two studies investigated the test–retest reliability spanning a period of a year. Stability coefficients for the scales ranged from .33 to .67 in a sample of 700 teachers (Jackson, Schwab, & Schuler, 1986) and from .34 to .62 in a sample of 46 human services professionals (Wade, Cooley, & Savicki, 1986). In all studies, emotional exhaustion appeared to be the most stable burnout dimension, whereas depersonalization was the least stable dimension. These high correlations suggest that burnout is a chronic rather than a transient state of mind.

Because validity is a generic concept, three aspects are briefly discussed: factorial validity, convergent validity, and discriminant validity. The *factorial validity* of the MBI is not completely beyond question. Although the three-dimensional structure of the MBI has been confirmed in several studies (Belcastro et al., 1983; Gold, 1984; Huberty & Huebner, 1988; Fimian & Blanton, 1987; Fimian et al., 1989; Green & Walkey, 1988; Koeske & Koeske, 1989; Lahoz & Mason, 1989; Pierce & Molloy, 1989), other researchers have found two dimensions (Brookings et al., 1985; Dignam, Barrera, & West, 1986; Green, Walkey, & Taylor, 1991), and even four dimensions (Firth, McIntee, McKeown, & Britton, 1985; Iwanicki & Schwab, 1981; Powers & Gose, 1986).

Thus far, virtually no adequate attempts have been made to test the factorial validity of the MBI, by carrying out confirmatory factor analysis. Although Golembiewski and Munzenrider (1988, pp. 19–23) claim that the results of their confirmatory factor analyses support the three-factor structure of the MBI, they provide insufficient information about the rather unusual method they used, i.e., Ahmavaara's technique (see Chapter 14). Moreover, Golembiewski and Munzenrider employed a strongly modified version of the MBI. Four recent studies examined the dimensionality through confirmatory factor analysis with LISREL and found the fit of the original three-factor model to be superior to several alternative models (Gold, et al., 1989; Lee & Ashforth, 1990; Byrne, Schaufeli, & Van Dierendonck, 1992).

Most studies on the *convergent validity* of the MBI have been discussed above. Generally, these studies yield positive results, indicating that to a certain extent the MBI scales measure the same construct as do other burnout instruments such as the BM, PBI, SBS-HP, and MBA. Roughly speaking, these measures share about 25% of their variance. As was concluded before, emotional exhaustion is the best validated dimension of the burnout syndrome. Unfortunately, the relationships with self-ratings, peer ratings, and expert ratings of MBI burnout are less strong (Meier, 1984; Rafferty et al., 1986). For instance, correlations between peer ratings of burnout and MBI scale scores range from .20 to .56 in various samples of social service workers ($N = 91$), physicians ($N = 43$), nurses and social service workers ($N = 180$), and police officers and their spouses ($N = 142$) (Maslach & Jackson, 1986). Generally, only about 10% of the variance of the MBI is shared with information from external sources, and again emotional exhaustion appears to be the best validated MBI dimension.

There is an evident discrepancy between these results and those found in validation studies that rely exclusively on self-report measures. It is likely that at least some portion of the common variance between self-report measures of burnout is caused by method variance and can be considered an artifact. Thus we are left with the conclusion that, except for the emotional exhaustion scale, the convergent validity of the MBI has not yet been demonstrated convincingly.

Well-designed investigations that assess the *discriminant validity* of the MBI are very rare. Employing a multitrait-multimethod approach, Meier (1984) concluded from a study of 320 faculty members that considerable overlap exists between the MBI and several measures of depression. Unfortunately, he used a total MBI score that summed responses across all items (an unorthodox procedure that was criticized by the test authors). However, Firth et al. (1986b) scored the MBI in the usual way and arrived at an interesting conclusion in their study of 200 nurses. Emotional exhaustion was substantially related to depression ($r = .50$), as measured by Beck's Depression Inventory. Relationships with personal accomplishment ($r = -.17$) and with depersonalization ($r = .32$) were much lower. Similar results were obtained by Landsbergis (1988), who used a less well-known depression scale.

Although in the MBI test manual only weak and insignificant relationships with job satisfaction are reported, several studies show much stronger associations (e.g., Dolan, 1987; Eisenstat & Felner, 1984; Koeske & Koeske, 1989; Landsbergis, 1988; Lindquist & Whitehead, 1986; Penn, Romano, & Foat, 1988; Stout & Williams, 1983). The results from these studies are somewhat consistent, indicating that emotional exhaustion has moderately negative correlations with job satisfaction (coefficients ranging between .35 and .45). Depersonalization is only slightly negatively correlated (coefficients ranging between .25 and .35), whereas personal accomplishment is positively but insignificantly related to job satisfaction.

Finally, some remarks have to be made about the classification of levels of burnout as presented in the test manual. This classification is based on arbitrary statistical norms. The normative sample has been arbitrarily divided into three equally sized groups of 33.3%, assuming that one third of the subjects will experience a high degree of burnout, another third will experience an average level of burnout, and the final third will experience a low level of burnout. However, as Einsiedel and Tully (1982) indicated, no clinically valid reason exists for using the top-third/bottom-third split as the dividing line instead of using, say, the median or the mean. Neither of these cutoff points has been empirically validated, by employing independent expert ratings from clinical psychologists or psychiatrists.

Maslach and Jackson note, correctly, that the categorization of burnout into three levels "is intended primarily as feedback for individual respondents" (1986, p. 5). Moreover, they explicitly warn that "neither the coding nor the original numerical scores should be used for diagnostic purposes; there is insufficient research on the pattern(s) of scores as indicators of individual dysfunction or the need for intervention." Researchers are directed to use the original numerical scores rather than the categorizations of low, average, and high for statistical analyses. Despite this clear statement, there is a tendency for some researchers to use a particular cutoff point (usually the top third) to differentiate burnout cases from noncases (e.g., Ackerley et al., 1988; Belcastro et al., 1983; Birch, Marchant, & Smith, 1986; Firth et al., 1985; Firth & Britton, 1989; Lindquist & Whitehead, 1986; McGrath, Reid, & Boor, 1989; Lahoz & Mason, 1989; Penn et al., 1988; Turnipseed, 1987; Ursprung, 1986). They then compute the proportion of burned-out persons in their particular sample. With only one exception (Ackerley et al., 1988), less than one third of the respondents are classified as high in emotional exhaustion or in depersonalization. Obviously, the subject populations do not represent the full range of those experiencing burnout, and they have other more skewed distributions than does the normative sample. This is particularly true for personal accomplishment. The percentage of respondents classified as low on this burnout dimension ranges between 1% (Ackerley et al., 1988) and 97% (McGrath et al., 1989)! Such remarkable differences underscore the argument of the test authors not to use the burnout categories but to look at the actual scores instead. Furthermore, no empirical basis exists yet for employing the MBI as a diagnostic tool for individual assessment.

The conclusion concerning the psychometric quality of the MBI is somewhat inconsistent. On the one hand, the factorial and the convergent validity as well as the reliability of the instrument is quite encouraging. On the other hand, burnout as measured with the MBI cannot be validly distinguished from related concepts such as depression and (to a somewhat lesser degree) job satisfaction. It is noteworthy that the scale that is most robust and reliable and that displays the strongest convergent validity (i.e., emotional exhaustion) is the least specific dimension of burnout, showing considerable overlap with related constructs.

CONCLUSIONS AND RECOMMENDATIONS

Our review of the literature shows that during the past decade, much progress has been made in the measurement of burnout. In this closing section, seven tentative conclusions are drawn.

1. The vast majority of instruments that assess burnout are self-report measures. This causes a particular problem in validation studies that use self-

report measures exclusively because at least part of the common variance of the measures has to be attributed to method variance.

2. Most instruments are designed to assess levels of burnout in human services professions. They should not be applied in other occupational contexts because it cannot be assumed that the structure of the burnout syndrome is identical across different occupational groups (see Chapter 1).

3. Two self-report instruments have been intensively studied psychometrically (i.e., BM and MBI). Unfortunately, little information is available about the psychometric properties of the remaining self-report instruments. Mostly, their authors only report satisfactory internal consistencies.

4. Although the psychometric qualities of the BM are promising, its use is rather limited because of the unidimensionality of the measure. The BM reduces a complex psychological phenomenon to mere exhaustion.

5. The psychometric qualities of the MBI (the most widely used self-report instrument) are encouraging but not completely beyond question. In particular, its discriminant validity is rather poor.

6. In most self-report inventories the individual's depletion of emotional resources is included in one way or another. Paradoxically, this crucial dimension of burnout also seems to be its least specific component (see also Chapter 2).

7. Despite the obvious agreement on the core meaning of burnout (i.e., exhaustion of a person's resources), considerable confusion exists about the number, as well as the nature, of other dimensions involved. For instance, some instruments not only assess burnout on the individual level but include organizational aspects as well.

DIRECTIONS FOR FUTURE RESEARCH

In the remainder of this chapter six directions for future research are discussed.

Development of Alternative Burnout Instruments

Currently, burnout is almost exclusively assessed with self-report questionnaires. Occasionally, self-assessment with rating scales or peer ratings is employed. Additionally, standardized interviews should be developed as well as behaviorally anchored rating scales that can be assessed by supervisors and colleagues. Such alternative assessment methods can be cross-validated with the MBI because this currently seems to be the most promising self-report questionnaire to measure burnout. Moreover, specific burnout instruments have to be developed that can be applied outside the human services professions. This means that, in particular, the depersonalization dimension has to be reconsidered. Basically, depersonalization is a specific form of psychological withdrawal from the essence of one's job. In human services professions, recipients constitute the core element of the job. In management, however, the core element is the organization. It is likely that burned-out managers withdraw psychologically from the organization by developing negative attitudes and behaviors toward the organization and its members. Therefore, a burnout measure for managers should include such particular attitudes and behaviors.

Psychometric Improvement of the MBI

The MBI can be improved in a number of ways. Additional psychometric development of the depersonalization dimension seems necessary. Depersonalization is the shortest (and therefore least reliable) scale of the MBI, with the most complex factor loadings. Adding a couple of items, preferably regarding the behavioral element of depersonalization, would not only increase the internal consistency (cf. Golembiewski & Munzenrider, 1988, pp. 19–21) but also strengthen the validity of this scale. Moreover, an equal balance of positively and negatively worded items is strongly recommended. In the present version, the items of the emotional exhaustion and depersonalization scales are phrased negatively, whereas those of the personal accomplishment scale are phrased positively. It cannot be completely ruled out that this fact explains the substantial correlation between the first two scales. Finally, factorial validity studies suggest that one item ("I feel energetic") should be deleted from the MBI since it not only loads on the intended personal accomplishment dimension but also on emotional exhaustion (Koeske & Koeske, 1989; Mor & Laliberte, 1984; Schaufeli & Van Dierendonck, 1992). Inspection of the MBI manual (Maslach & Jackson, 1986, p. 30) reveals that this item is the weakest and most complex item in the accomplishment scale.

Individual Assessment of Burnout

A great need exists for instruments that can assess burnout on an individual level. The MBI could serve this purpose, but the existing categorization into upper, middle, and lower thirds is arbitrary and lacks any clinical validity. Therefore, this categorization is inappropriate for individual assessment (as is rightfully stressed in the test manual). What is needed are studies that establish clinically valid cutoff points by comparing burnout scores with expert ratings of psychiatrists or psychologists. Such a validation procedure can be difficult, as demonstrated by the discriminant validity study of Rafferty et al. (1986) that was reviewed in the first part of this chapter. The study found rather weak relationships between the MBI scores and expert ratings, possibly because the researchers did not clearly specify the burnout syndrome to the experts. Furthermore, appropriate group norms should be developed based on representative, stratified random samples of different professions. Many researchers employ the "thirds" categorization of MBI scores in order to estimate the number of burned-out cases in a particular profession. Unfortunately, not only is the use of this categorization invalid, as was previously shown, but the samples are usually inappropriate. Only the few studies that employ appropriate samples, allow for generalizations to the professional group as a whole and thus can be used to develop valid group norms. These notable examples are studies of social workers (Himle, Jayaratne, & Thyness, 1986), prison guards (Lindquist & Whitehead, 1986), librarians (Birch et al., 1986), and psychologists (Ackerley et al., 1988).

Epidemiological Research

In developing adequate group norms along the lines outlined above, one actually studies the epidemiology of burnout in different professions. With the MBI we have a reasonably reliable and valid instrument to assess burnout in human services professions. There is a great need for epidemiological knowledge of burnout in order to identify specific (sub)groups at risk. For instance, virtually no studies exist that provide the sort of information that would be necessary for planning appropriate interventions. To date, most epidemiological studies in the occupational field have used the General Health Questionnaire (Goldberg, 1978) to assess minor psychiatric disorder. However, this instrument has been criticized because it is somewhat unspecific (Fletcher, 1988; Warr, 1990). The advantage of the MBI is its twofold specificity concerning the domain (three aspects of burnout) as well as the occupational group (human services professions).

Cross-National Psychometric Research

The growing popularity of the MBI outside English-speaking countries requires a thorough psychometric evaluation of this instrument in each specific national context. With only two exceptions (Schaufeli & Van Dierendonck, 1992; Schaufeli & Janczur, in press), such studies have only been published in local languages (e.g., Enzmann & Kleiber, 1989; Gil-Monte & Schaufeli, 1992; Girault, 1989; Sirigatti, Stefanile, & Menoni, 1988). Although the results in Chapter 13 suggest that the MBI can be used cross-nationally, valid research that compares levels of burnout between countries is virtually lacking. To date no attempts have been made to calibrate the MBI for different nations, yet such calibration is a prerequisite for using the instrument for diagnostic purposes.

Development of Conceptual Models of Burnout

Finally, and most important, research with the MBI should generate and test models about the etiology and the persistence of burnout in the human services. Strictly speaking, this is not a psychometric endeavor; as was pointed out before, assessment of the discriminant validity of the MBI is only possible within an elaborate theoretical framework. Because of its multidimensional conceptualization of burnout, the MBI seems to be particularly appropriate to employ in theory-driven research. The fruitfulness of such an approach is demonstrated by several authors who contributed to this volume (Buunk & Schaufeli, Golembiewski et al., Maslach, and Leiter). We strongly believe that, despite its minor weaknesses, the MBI has a promising future in extending our knowledge of professional burnout.

13

BURNOUT IN CROSS-NATIONAL SETTINGS: GENERIC AND MODEL-SPECIFIC PERSPECTIVES

Robert T. Golembiewski and Katherine Scherb
University of Georgia

Robert A. Boudreau
University of Lethbridge

Even if the conclusion must be clearly provisional, mounting evidence implies that burnout is a cross-national and perhaps a cross-cultural disease, as this brief survey illustrates under two broad headings. Under a generic rubric, a variety of perspectives and ways of estimating burnout are marshaled to support this general conclusion. Under a model-specific rubric, research with the "phases of burnout" using a single operational definition also supports the present conclusion that burnout is not strictly culture-bounded.

As far as this chapter has an overarching framework or organizing principle, it is a simple if significant one. What evidence or clues does a comprehensive review of the available literature reveal about the culturally loaded or cross-national character of burnout? Thus, the objective is an expansive one even though the products of analysis cannot rise above the available literature, which has only fairly begun to exploit cross-cultural perspectives.

Four emphases constitute the present contribution to exploring our macro focus. First, a conventional review of the cross-national literature provides overall perspective. Second, evidence about cross-national congruence of the three subdomains of the Maslach Burnout Inventory (MBI; Maslach & Jackson, 1986) gets detailed attention, along with notes on conceptual diffusion and penetration of the burnout concept. These we consider generic perspectives on whether burnout is cross-national or culturally loaded. Third, a review of model-specific evidence follows, with the target being the phase model of burnout (Golembiewski, Munzenrider, & Stevenson, 1986; Golembiewski & Munzenrider, 1988). Fourth, some general conclusions are drawn in a closing section.

Two caveats require early note. In both the generic and model-specific emphases of this paper, the focus is on cross-national samples. Arguably, these neither clearly nor necessarily constitute cross-cultural comparisons. For present purposes, however, the distinction gets no further attention despite its patent significance.

Moreover, as noted, the available evidence inclines toward a conclusion that burnout is cross-national and perhaps cross-cultural, despite clear inadequacies and lacunae. A conclusion emphasizing national—or cultural—boundedness would grate less on today's academic nerve-endings. Witness this contrary conclusion (Adler, Doktor, & Redding, 1986, p. 295): "Research in developmental psychology, sociology, and anthropology shows that there are major differences among the cognitive processes of people from different cultures. In the era of the global corporation, cultural diversity has to be recognized, understood, and appropriately used in organizations." So the present broad conclusion must swim against a substantial tide, as it were.

EMERGING EMPIRICAL RESEARCH

A conventional survey of the burnout literature rests on a benchmark bibliography of over 700 items compiled by Kilpatrick (1986), whose cross-national entries were augmented by a restricted search of English language sources extending her coverage from 1985 through the end of 1990. This augmentation relied on computer-assisted searches of the periodic literature using a range of key words and concepts; visual searches of *Psychological Abstracts* and *Sociological Abstracts* as well as of numerous journals; and personal contacts. This multipronged search isolated a panel of 28 citations that can be reasonably labeled "empirical and cross-national." The items are listed in the References, and each carries an identifying asterisk.

The cross-national panel encourages nine generalizations about the available literature on burnout-related phenomena, as described in the following sections. Several generalizations may be considered robust, but most require qualifications about their tentative and provisional character, especially since there are so many operational definitions of "burnout" or stress-related phenomena.

Cross-National Research's Growth Curve

Only rare cross-national efforts with even a conceptual kinship to burnout appear before the early 1980s. Beyond that, two regularities seem clear. Thus the incidence of burnout-related research in cross-national settings increased markedly, as has the specific focus on burnout rather than on stress or stressors.

Cross-National Research and Self-Reports

Virtually all studies in the present panel depend on self-reports, which estimate the levels, sources, and mediators of stressors or strain, as diversely defined (see "Cross-National Research and Operational Measures" on p. 225). In only a minority of cases, moreover, is the focus explicitly on burnout, which can be broadly defined as "strain" or levels of stressors that surpass an individual's comfortable coping capacities—as determined by attitudes, expectations, and coping skills.

Multiple "hard" and "soft" measures in every study no doubt constitute the ideal, and hence most observers would entertain concerns about the reliability and validity of available cross-national research. Some observers would insist on hard measures as alone capable of generating reliable and valid results.

Cross-National Research and "Meaning"

In at least two senses, the focus on "meaning" gets too little attention. Thus work itself may have different meaning(s) in cross-national or cross-cultural settings, and only promising starts exist here (e.g., England, 1988; England & Whitely, in press).

At a micro level, instruments used to assess opinions and attitudes via self-reports often pose their own issues of meaning in cross-national applications. This area remains basically an art form, building on such useful procedures as translation followed by back-translation. As for the MBI translations, they seem to have encountered no insurmountable problems (e.g., Boudreau & Golembiewski, 1989, 1990).

Cross-National Research and Gross Validity

Especially in earlier years, burnout-related research seemed preoccupied with establishing its *raison d'etre*, which typically was reflected in efforts to associate the kind and intensity of stressors with noteworthy health outcomes. Such studies report clusters of effects that seem common cross-nationally as well as others that seem setting-specific. For example, consider a few selected findings from one of the earliest cross-national studies (Romo, Siltanen, Theorell, & Rahe, 1974). Two hundred and twenty-nine men from Finland, Sweden, and the United States who survived a myocardial infarction completed a questionnaire soliciting self-reports about work behavior, time urgency, and life dissatisfactions. Six of the 14 questions had high subscription rates by men in all three samples: responsibility at work, time urgency, hostility when slowed by others, overtime work, dissatisfaction with level of education, and dissatisfaction with achievement of life goals. However, both Scandinavian groups expressed more job dissatisfaction than the Americans. The Finnish sample registered nearly twice the level of life dissatisfactions as the Swedes and Americans, perhaps because of Finland's relatively disadvantaged socioeconomic condition.

Other early cross-national research similarly concluded that stress-related research isolated major health covariates. For example, Orth-Gomer (1979) focused on ischemic heart disease (IHD) among matched Swedish and American white men, with each sample having three subgroups: men registered with the medical departments of their companies for IHD (myocardial infarction or angina pectoris); men with one or more indicators of strong IHD risk; and a control group. A questionnaire administered to all the subjects in both samples inquired about job satisfaction, conflicts, psychological strain, and stressful periods. Significant differences were observed between samples. Swedish men in all three subgroups were less satisfied with their careers than their American counterparts, while no consistent differences existed in career satisfaction between subgroups within each sample. Men with IHD reported the most overtime and highest job pressure in the Swedish sample, but, in contrast, the control group reported the most overtime and job pressure in the American group. Also suggestive of national differences, Swedes—and especially those with IHD—associated peak stressfulness almost exclusively with problems at work. Americans—mainly those in the IHD and IHD-risk subgroups—saw family problems as the closest precursors of peak stressfulness. Noteworthy similarities also were observed. Paramountly, both Swedish and American samples reported that very stressful periods preceded IHD onset. In both samples, moreover, men with IHD had less education and experienced significantly less satisfaction with family life.

This brief review focuses on "gross validity." That is, no cross-cultural studies targeting burnout and health-related outcomes have been conducted, although results from separate studies in the United States (Golembiewski & Munzenrider, 1988, pp. 68–77) and Japan (Golembiewski, Boudreau, Goto, & Murai, 1992) generate significant and similar patterns of association, using the same operational measure of burnout. The review above concentrates on findings from studies featuring concepts related to burnout.

Cross-National Research and Work/Job

Cross-national comparisons moved closer to explicit burnout formulations when they involved work or job strain. Illustratively, Karasek (1979) used data from national surveys in the United States and Sweden to test a model of job strain viewed as deriving from the mental strain inherent in the interaction of job demands and job decision latitude. For both countries, paramountly, workers with jobs simultaneously low in job decision latitude and high in job demands reported exhaustion after work, depression, nervousness, anxiety, insomnia, and trouble awakening in the morning. Data on pill consumption and sick days were available for the Swedish sample, and jobs with low decision latitude and high demand were strongly associated with pill consumption and use of sick days. Signally, the combination of low decision latitude and high demand was also associated with job dissatisfaction. Both the satisfaction measures and the depression indicators showed some covariation with the activity level of the job. Active jobs with high demands and high decision latitude were more satisfying than passive jobs. The latter findings imply a crucial suggestion, consistent with other findings (e.g., Golembiewski, Hilles, & Daly, 1987). Job strain might be reducible by increasing decision latitude, independent of changes in workload. This suggests an alternative to a troubling forecast: that job-related health can be improved only by decreasing productivity.

Cross-National Research and Similarities/Differences

Viewed from another angle, the panel of cross-national studies reflects several other interesting points of contrast and consistency. Five summary points illustrate clusters of these similarities/differences, and they also serve to amplify and elaborate the rationale for studying burnout-related phenomena.

First, at the broadest level, research suggests that strain-related—and hence burnout-related—effects are of international interest, and nowhere more so than in the "developing" or rapidly industrializing countries where effects seem especially virulent. Consider the 1,065 executives in the Cooper and Arbose (1984) study (see also Cooper & Hensman, 1985). Those researchers report that respondents who registered the highest mental ill health and lowest job satisfaction scores came from the developing countries—Singapore, Nigeria, Brazil, Egypt—as well as from one advanced country: Japan. For executives from these countries, significant stressors include work overload, time pressures and deadlines, long work hours, taking work home, and poor interpersonal relationships. In the more advanced or mature nations—Britain, United States, South Africa, Sweden, and Germany—threat of job loss and lack of autonomy were predictors of mental ill health and job dissatisfaction. Idiosyncratic predictors of stressors also characterized specific countries. Witness the competition for promotion and keeping up with new technology cited by the Japanese respondents.

Second, replications tend to find high levels of stressors and poorer mental well-being among similar clusters of countries. For example, consider comparisons between American, Japanese, and Indian managers (DeFrank, Ivancevich, & Schweiger, 1988). The Japanese subsample—as seems typically the case (e.g., Boudreau & Golembiewski, 1989, 1990)—registered higher levels of stressors and mental ill health. A principal components analysis of the mental health scores from each country produced a three-factor solution, with the factors designated as Tension, Dissatisfaction, and Ill Health. Overall, the Japanese reported poorer health status and significantly more tension than the American and Indian managers. Japanese and Americans did not differ on Dissatisfaction, but both groups were significantly higher than the Indian sample. Note also that distinguishing between sources of support helped clarify major relationships. Ill Health was negatively correlated with coworker support among the Japanese, negatively correlated with boss support among the Indians, and positively correlated with support from relatives among Americans.

DeFrank et al. (1988) also suggest an important methodological issue for cross-national work—indeed, for all comparative research. Their results highlight the importance of examining the factor structures of measures, in addition to relying on total scores. Specifically, in their Global Stress Scale, a Control factor accounted for the highest portion of variance among the Japanese and Indians, while a Time Concerns factor accounted for the highest portion of variance among American managers. Differences also appeared in the percentage of variance in mental health scores accounted for by the three factors.

Third, consistently, replications also tend to isolate more favorable burnout-related profiles in respondents from countries with a reputation for relaxed lifestyles. Illustratively (McCormick & Cooper, 1988), mental health, job satisfaction, and job stressors among senior executives in New Zealand were lower than in the 10 countries in Cooper and Hensman's (1985) study, excepting Sweden. As the researchers expected, the New Zealand executives reported the lowest rates of job dissatisfaction.

Fourth, the points above suggest that cross-national comparisons often tap pervasive cultural/behavioral differences, but that conclusion must be tempered. Consider here only the possibility that, for example, the specific features of *some* roles or tasks might distort or even wash out cultural/behavioral differences. For example, Shouksmith and Burrough (1988) compared self-reports from New Zealand air traffic controllers with responses to a set of items previously rated as stress producing by Canadian air traffic controllers. The overall level of stress perceived by the two groups was approximately equal, and four of the top five stressors identified by both groups were the same (equipment limitations, workload in peak traffic situations, fear of causing accidents, and poor quality of the general working environment). This suggests that much strain in air traffic controllership inheres in common role and task factors, as distinct from differences in the national or cultural settings.

A similar conclusion about role similarities in different national settings holds for organizational consultants from European countries, Israel, and the United States (Pines & Caspi, 1992). No cross-national differences in burnout were observed. Although only a small number of consultants were involved (*N* = 40), the study also reported familiar burnout covariates, which add to the study's credibility.

The impact of tasks/roles also may be moderated in an interesting way, as research with university faculty suggests. Keinan and Perlberg (1987) administered a questionnaire using the Faculty Stress Index (FSI) to members from all Israeli universities and compared the results to those of an earlier survey of American faculty (Gmelch, Lovrich, & Wilke, 1983). Factor analysis of the FSI items indicated five distinct factors that seem similar to those identified in the Gmelch study: conflicts with the academic system, overload of administrative and public duties, academic overload and time constraints, teaching functions, and working conditions. Seven of the top 10 stressors identified by the Americans and Israelis were common to both samples.

However, Keinan and Perlberg (1987) also showed that lower percentages of Israeli faculty perceived each source of stress as "serious" or "severe" than American faculty. Research, teaching, and service were all perceived as more stressful by Americans than by Israelis (see also the research discussed later by Etzion and her associates). In short, the *domains* may be cross-nationally similar while their *levels* may vary markedly.

Fifth, in some selected particulars, cross-national samples seem to generate similar consequences. Perhaps most confidence can be placed in gender, with virtually all studies implying higher levels of burnout-related effects among females. For example, Cohen (1976) reported that females in a sample of American retail clerks and German retail clerks and factory workers experienced higher rates of felt discomfort than their male counterparts. Similarly, in Keinan and Perlberg's (1987) survey of university faculties, female respondents reported higher levels of stressors. Consistently, Etzion and Pines (1986) found that the women in a sample of Israeli and American human service professionals reported significantly more burnout than did men.

Such apparent similarities concerning gender effects can be accounted for in terms of similar cross-national socialization and acculturation patterns. Thus, Etzion and Pines (1986) suggest the centrality of sex role stereotypes of masculine and feminine behavior, as well as of cultural values.

Cross-National Research and Personal Features

Little comparative research exists concerning the relationship of personal or personality features to stressors or burnout. For example, several studies look at measures of type A, but the results can be considered provisional and perhaps highly contingent (e.g., Evans, Palsane, & Carrere, 1987; Xie & Jamal, 1989).

Cross-National Research and Cultural Profiles

Several of the seven summaries above encourage the search for distinct "cultural profiles," to use a convenient term to denote a very complex set of realities. Such profiles permit explicit characterizations of the contexts inducing broad similarities/differences in research populations, and search for them may eventually generate broadly useful topologies.

Generally, such cultural profiles have proved useful in understanding the results of burnout-related research. For example, one of the few studies providing data from eastern Europe examined stress predictors and mediators among American and Polish college students (Harari, Jones, & Sęk, 1988). The American students scored higher on internal locus of control and social support, with the Poles scoring higher on external locus of control, anxiety, and depression. For the Americans, internal locus of control buffered anxiety and depression, consistent with the researchers' assumption that an individualistic-privatistic orientation dominated among Americans. However, in the Polish sample—to which the researchers attributed a collectivistic-institutional orientation—social support did *not* buffer stressors. Contemporary Polish conditions may explain this anomaly: low levels of self-confidence among highly educated Poles, as well as intense competition for resources in Polish society.

Much the same point about the relevance of cultural profiles, if in more detail and with sharper definition, derives from a series of studies on tedium and coping strategies among Israelis and Americans in a variety of occupations. Tedium is defined as the "experience of physical, emotional and mental exhaustion" (Etzion, Pines, & Kafry, 1983, p. 42), and consequently shares much conceptual ground with burnout generally, as well as with the MBI specifically (see Chapter 12). Note that Pines and Aronson (1988) more recently use the terms "burnout" and "tedium" in similar ways.

Overall, this line of research consistently finds lower levels of reported tedium among Israelis as compared to Americans (e.g., Etzion & Pines, 1986; Pines, Aronson, & Kafry, 1981; Pines, Kafry, & Etzion, 1980), and the character of the findings deserves note. For example, conflict between life and work was significantly higher for an American sample of managers (Etzion, Kafry, & Pines, 1982). Feelings of guilt and anxiety, as well as experiences of conflicting demands, emerged as significant positive tedium correlates for Israelis as well as Americans in both their life and work, while overload was a positive tedium correlate for both samples in life only. The research team places the templates of two cultural profiles over these results, as it were. First, Etzion and her associates (1982) suggest that strong social systems of family, friends, and neighbors help support Israelis in times of stress, whereas American culture emphasizes competition, individual achievement, and personal excellence. Moreover, due to the volatile political climate, limited geographic mobility, and small size of the country, Israelis have a greater sense of social unity and mutual fate. The researchers also speculate that the greater conflict between life and work among the American sample may be due to the importance of work as a source of social contacts.

Second, the Etzion research team in a similar way explains the consistent tedium associations with different cross-national preferences about coping strategies. In a study of Israeli and American human service professionals (Etzion et al., 1983), the latter reported using all the passive-direct and passive-indirect coping strategies, as well as the active-indirect technique of getting involved in other activities, more frequently than Israelis. In both samples, confronting the source of stress (an active-direct strategy) was a significant negative tedium correlate while the frequency of using two passive-indirect techniques (getting ill and collapsing) was a significant positive tedium correlate. Finding positive aspects in difficult situations also seemed a significant negative tedium correlate for Israelis only, while avoiding and leaving the source of stress were significant positive tedium correlates for the Americans only. Both samples consistently perceived the passive coping strategy as less successful in reducing stress than active techniques. For Americans, all correlations between tedium and success of the active coping techniques were negative and significant. For the Israelis, success of an active coping strategy almost always was unrelated to tedium. Again, these tendencies may reflect culturally determined values. Israelis apparently valued active modes of coping for their own sake, while Americans valued active techniques only when they proved successful.

Cross-National Research and Operational Measures

The numerous operational measures utilized in cross-national research on burnout constitute a major obstacle to the interpretation of the present panel of studies (see also Chapter 12). Thus most operational measures focus on *stressors*, as is true of the School Psychologists and Stress Inventory, or SPSI (Burden, 1988) and the lists associated with teaching (e.g., Tokar & Feitler, 1986), among many others. In contrast, a few operational measures direct attention at the resultant *strain* (e.g., Etzion et al., 1983; Maslach & Jackson, 1981a, 1986). Many other variants could be cited.

Overall little is known about the conceptual and mensural overlap of these several operational definitions, both within a nation or culture as well as between two or more of them. The single exception is the Maslach Burnout Inventory, or MBI (Maslach & Jackson, 1981a, 1986), as subsequent discussion establishes.

This hardiness of the MBI subdomains justifies special confidence in the results generated by research designs using this operational measure. Perhaps the most substantial cross-national study using the MBI involved over 1,000 Norwegian and U.S. social workers drawn from randomly selected national samples (Himle, Jayaratne, & Thyness, 1986), and its findings illustrate the typical covariates of the MBI subdomains (i.e., emotional exhaustion, depersonalization, and reduced personal accomplishment; see also below). In both samples, challenge of the job emerged as the most common and strongest predictor of all three subdomains of burnout, job satisfaction, and turnover. Value conflict was a significant predictor of depersonalization in the Norwegian sample and was associated with lower job satisfaction and turnover in the American group. Role conflict emerged as a significant predictor of emotional exhaustion and depersonalization among the Americans, but was associated with decreased job satisfaction for the Norwegians. Role ambiguity was inversely associated with personal accomplishment for the American sample but not for the Norwegians. In the American sample, age was inversely associated with emotional exhaustion and depersonalization and positively related to personal accomplishment, replicating Maslach and Jackson's (1981a) findings on the relationship between age and the MBI subdomains among U.S. social service workers. In contrast to Maslach's (1976) finding that workload is positively related to burnout among human service workers, however, workload was not a significant predictor in the study by Himle et al. (1986).

A few researchers have devoted attention to the comparison of different operational definitions. For example, Schaufeli and Van Dierendonck (1992) compared the MBI and the Burnout Measure (Pines & Aronson, 1988), which was originally represented as a measure of "tedium" (Pines et al., 1981). The researchers found some overlap between the measures, but also differences that might be expected to generate diverse patterns of findings.

However, in the general absence of similar comparative work on other operational definitions, summaries of burnout-related research have to be tentative. To be sure, the literature summarized above clearly has certain central tendencies, but far greater specificity is required. For example, consider studies reporting differences/similarities in covariates generated by research designs using different operational definitions, e.g., concerning "stress levels" in cross-national school settings (e.g., Burden, 1988; Manso-Pinto, 1990; Tokar & Feitler, 1986).

How to interpret those differences/similarities? One might argue that similarities in covariates reflect a burnout-related centroid in nature, if you will, that is pervasive enough to be isolated by even very different operational measures of "stress" or "strain" or "burnout." But what of the differences in covariates? They might result from differences in operational definitions, e.g., *stressor*-based measures versus those tapping *strain*.

This overview cannot solve such crucial problems related to operational definition, but it endeavors to be sensitive to them even as it deliberately elects at this early stage not to be highly restrictive about what qualifies as *the* operational definition(s) of burnout. Hence in the discussion above, "burnout-related" or "stress-related" are circumlocutions that intentionally designate the targeted phenomena in generous terms. This convenience stands as an uneasy surrogate for the unavailable specificity about the numerous operational definitions used in our cross-national panel of burnout studies.

SOME GENERIC EVIDENCE

A second perspective on the accumulating attention to burnout as cross-cultural or cross-national is here called generic to distinguish it from a second following major section that is model-specific. The dual aspects here dominating the generic view give attention in turn to the deep penetration and broad diffusion of burnout as concept, and the general as well as growing agreement as to burnout's component subdomains.

First, in both popular and research circles, the burnout concept has spread both wide and deep in a few years. A broad spectrum of evidence could be marshaled in support of this view, but space permits notice here only that several chapters in this volume provide such evidence (e.g., Chapters 1, 5, and 6).

Second, in a critical methodological sense, growing evidence suggests that burnout can be assessed in similar terms—both within and between large sociopolitical aggregates. Consider the statistical treatment of the items of the MBI (Maslach & Jackson, 1981a, 1986), which is by far the most commonly used instrument (Kilpatrick, 1986). The MBI items tap these three conceptual subdomains (see also Chapter 2):

1. Depersonalization, on which high scores identify individuals who distance themselves from others, and who tend to reify their human relations
2. Personal accomplishment (reversed), on which low scores refer to individuals who believe they are doing well on a task that they consider worthwhile
3. Emotional exhaustion, on which high scores indicate persons who are experiencing strain at or beyond their normal coping limits, who in the vernacular are "at the end of their rope"

A key issue associated with the MBI items—perhaps *the* key issue—involves the question of whether separate studies assess the same dimensions of reality in different national samples. In order to study this issue, responses from several national samples are factor-analyzed (Principal Factors Analysis followed by varimax rotation toward three factors). U.S. and Canadian employees responded to an English version of the MBI format. Samples from four other nation states—China, Japan, Poland, and Yugoslavia—responded to translations of the MBI items by social scientists with appropriate first languages. Translations did not pose great problems but some adjustments were required. For example, the Japanese language has no direct equivalent of the usage "being at the end of one's rope," which appears in an MBI item.

Further, this study employs Ahmavaara's (1954) method, which produces two measures of the congruence of each pair of factorial structures:

1. A product-moment coefficient, which estimates the congruence of "patterning" or profile of any two factorial structures.
2. An intraclass coefficient, which compares two structures not only with respect to patterning but also relative to "magnitude." This coefficient estimates the degree to which the loadings in one matrix are higher or lower than in the other, as well as assesses their similarity of profile.

The present criterion of "similarity" is conservative: estimates of congruence in the mid-80s indicate "similar" structures, and the two Ahmavaara coefficients are best read, when squared, as estimates of the variance shared by two factorial structures. Note also that the basic inputs to Ahmavaara's procedure are correlation matrices from two factorial analyses. Hence the number of subjects does not directly enter into the interpretation of the two primary Ahmavaara outputs although N will be a factor in judging the stability of a factor analysis.

Congruence of MBI Subdomains in U.S. Settings

As Kim establishes (1990, pp. 26–29), virtually all investigations with U.S. samples support the validity of a three-domain model of burnout, even when different methods of analysis are employed. Broadly, numerous studies of concurrent validity highlight the usefulness of the three MBI subdomains, (e.g., Golembiewski, Munzenrider, & Stevenson, 1986; Golembiewski & Munzenrider, 1988), although a few propose that depersonalization is dispensable (e.g., Gaines & Jermier, 1983). Correlational and factor analytic approaches (e.g., Golembiewski & Munzenrider, 1988, pp. 22–23; Kim, 1990, pp. 77–103) generate a similar conclusion on balance.

The use of Ahmavaara's technique permits a far more direct estimate of the specific magnitude of congruence. The earliest such U.S. comparison involves Maslach's original aggregate of nearly 2,000 respondents (Maslach & Jackson, 1981a) along with another U.S. sample of over 1,500 employees from about 50 regional offices of a federal agency identified throughout this chapter as "US 6." There, the congruence of the two independent factorial structures is substantial: .86 for the product-moment coefficient as well as for the intraclass coefficient. Subsequent analyses with other U.S. samples yield similar results (see Golembiewski & Munzenrider, 1988, pp. 18–23). Notably also, the U.S. results seem to closely parallel studies in other national settings (e.g., Schaufeli & van Dierendonck, 1992).

Congruence of MBI Subdomains in Cross-National Samples

Several cross-national comparisons also suggest that respondents to the MBI "see" substantially the same dimensional universe despite the underlying differences between national loci, work settings, and probably cultures. Here analysis compares samples from six countries with one another.

Table 13-1 provides substantial (if not comprehensive) perspective on the cross-national congruence of responses to the MBI items, summarizing as it does the results of Ahmavaara's analysis applied to all possible pairs of factorial structures from six national samples. In the table, the intraclass correlation coefficients approach a mean of .79. Perhaps significantly, in general, the larger the samples compared, the higher the correlations. US 6 has the largest sample, and presumably its factor structure is the most stable. The product-moment correlations are quite similar and, although they are not reproduced in the table, they usefully reject the possibility of noteworthy differences between the patterns and magnitudes of the factorial structures. Switching which nation in a pair is the "target" and which is the "problem" requires only minor qualifications of the results summarized above. Details do not get reported here due to space limitations.

Significantly, this evidence of cross-national congruence of the MBI items does not stand alone. Independent replications exist. Kim (1990) provides an overview of research supporting the present conclusion, and numerous separate demonstrations exist of the congruence of MBI subdomains (e.g., Green & Walkey, 1988).

SOME MODEL-SPECIFIC EVIDENCE

Limited but useful perspective on burnout as cross-national or cross-cultural derives from the phase model. Four emphases introduce the phase model's relevance while also reflecting its character and development. These emphases

TABLE 13-1 Cross-National Congruence of MBI Factorial Structures (Intraclass Correlation Coefficients)

	Target population				
	US 6 ($N = 1535$)	Canada 3 ($N = 399$)	China 1 ($N = 196$)	Japan 1–6 ($N = 914$)	Poland 1–2 ($N = 181$)
Canada 3	.88	—			
China 1	.83	.71	—		
Japan 1–6	.84	.81	.67	—	
Poland 1	.90	.83	.76	.80	—
Yugoslavia 1 ($N = 100$)	.79	.80	.71	.73	.72

include a sketch of the model's properties, a review of some major covariates of the phases, a selective review of the cross-national severity and incidence of burnout as operationally defined by the phases, and a brief summary of preliminary findings from a Japanese replication of the pattern of covariates associated with the phase model in North American work settings.

Phase Model of Burnout

Basically, the phase model assumes that the three MBI subdomains are differentially virulent from depersonalization to emotional exhaustion. Moreover, high/low norms for each subdomain derived from a large sample (Golembiewski & Munzenrider, 1988) permit three assignments for each individual. Following the single-decision rule about the progressive virulence of the subdomains, the high/low distinctions permit the generation of eight phases of burnout (see Table 13-2).

Beyond the MBI and its conceptual extensions, the phase model admits two basic forms of onset. *Chronic onset* refers to progressively worsening conditions at the worksite, or a kind of Chinese water torture approach to escalating burnout. Here, the single flight path is I → II → IV → VIII. *Acute onset* refers to some sudden, traumatic stimulus: e.g., the death of a loved one might precipitate a I or II into V or VI, respectively, with a difficult grieving process perhaps eventuating in a VII or VIII. Many other acute flight paths also are possible, depending on the character of the inducing stimulus (e.g., Golembiewski & Munzenrider, 1988, pp. 174–179).

Covariates of Burnout

Since the phases propose progressive virulence, individuals in advanced phases will experience more serious consequences than those in less advanced phases. However, no individual will proceed through each of the eight phases on the way to full-term burnout. For example, psychologically, it seems awkward for a II to move to a III.

The expectation that the phases I → VIII map worsening covariates gets

TABLE 13-2   Phases of Burnout

MBI subscale	I	II	III	IV	V	VI	VII	VIII
Depersonalization	Lo	Hi	Lo	Hi	Lo	Hi	Lo	Hi
Personal accomplishment (reversed)	Lo	Lo	Hi	Hi	Lo	Lo	Hi	Hi
Emotional exhaustion	Lo	Lo	Lo	Lo	Hi	Hi	Hi	Hi

Note: Lo = low; Hi = high.

substantial support. Some 300 variables have been tested for association with the phases, and in almost all cases covariates worsen as the phases progress. For example, phase by phase (e.g., Burke & Deszca, 1986; Golembiewski & Munzenrider, 1988), individuals on average will experience

- Decreasing satisfaction with work
- Heightening tension at work
- Poorer performance appraisals
- Decreasing self-esteem
- Higher turnover
- Greater physical symptoms
- Greater negative affects such as anxiety and depression
- Higher rates of nonpsychotic psychiatric symptoms

Commonly, 90% or more of the cases reveal progressively worse conditions, phase by phase. More impressively, large proportions of the paired comparisons also attain statistical significance as well as falling in the expected direction (Golembiewski & Munzenrider, 1988).

Estimating Cross-National Incidence/Severity of Burnout

Since phase assignments in effect estimate how many people are experiencing which degree of personal burnout, distributions of the phases in national samples provide a sense of the epidemiology of burnout. As Table 13-3 shows, the best estimates now available suggest that burnout can lay a strong, if preliminary, claim to status as an international disease.

Three emphases dominate in Table 13-3, as a more detailed analysis establishes (Golembiewski, Boudreau, Kim, Munzenrider, & Park, 1990). First, the phases tend toward a bimodal distribution, with phases I and VIII accounting for the largest proportions of cases. This is a familiar pattern in the panel of over 15,000 U.S. cases (e.g., Golembiewski & Munzenrider, 1988, pp. 113–125) and distinguishes the phases from most behavioral measures, which have broadly normal distributions.

Second, despite obvious qualifications, the several countries have different incidences of the advanced phases. Canada has the fewest advanced cases, in general; the U.S. is intermediate; and Japan as well as China—by far—have the worst distribution of phases. Japan 1–6 respondents average 73.2% in the three most extreme phases, to be specific;* the six U.S. settings place 46.3% in the

---

*The Japanese distributions of phases in Table 13-3 heavily load the advanced phases. Table 13-4 consistently shows that two other substantial Japanese samples reflect similar distributions (see also Boudreau & Golembiewski, 1989, 1990).

TABLE 13-3 Phases of Burnout in Selected Cross-National Populations (Percentages)

National settings	Combined N	Phases of burnout (%)							
		I	II	III	IV	V	VI	VII	VIII
Canada 1-3	839	25.1	9.1	18.5	8.7	6.2	9.2	6.1	17.2
			52.7%		14.9%			32.5%	
China 1	196	10.2	6.6	3.6	8.7	3.6	25.5	4.6	37.2
			20.4%		12.3%			67.3%	
Japan 1-6	914	2.7	2.1	5.4	15.1	1.5	5.8	4.5	62.9
			10.2%		16.6%			73.2%	
Poland 1-2	181	16.6	8.3	14.9	7.7	8.8	9.4	9.4	24.9
			39.8%		16.5%			43.7%	
US 1-6	8270	25.1	3.7	10.3	1.3	3.4	15.4	10.0	20.9
			39.1%		14.7%			46.3%	
Yugoslavia 1	100	10.0	11.0	20.0	5.0	12.0	17.0	4.0	21.0
			41.0%		17.0%			42.0%	

Note: Data are contributed by R. Wayne Boss, Robert Boudreau, Allan Cahoon, Gloria Deckard, Robert T. Golembiewski, Ewa Maslyk-Musial, Robert F. Munzenrider, Luke Novelli, Josip Obradovic, Se-Jeong Park, Benjamin H. Rountree, Julie Rowney, and Jerry G. Stevenson. The national identifiers indicate the number of separate work settings: e.g., Canada 1–3's 839 cases come from three settings. Most cases come from white collar, managerial, and people-helping settings such as hospitals.

TABLE 13-4   Phases of Burnout in Two Japanese Samples by Percentages

N	I	II	III	IV	V	VI	VII	VIII
503	3.8	4.5	7.0	17.1	4.2	8.9	4.8	48.1
387	2.3	1.6	3.4	4.1	1.6	15.5	2.8	68.2

Note: Data from Boudreau & Golembiewski, 1989, 1990.

same phases; and the three Canadian settings generate 36.4% in phases VI–VIII.

Third, the Canadian and U.S. estimates may be conservative. Health care settings dominate in Table 13-3 for these two countries, and neither set contains data from large urban health care settings that serve impoverished patients and might well induce greater proportions of health care providers with advanced burnout. So the Canadian and U.S. data may underestimate the magnitude of their burnout problem.

Japanese Replication of Covariates of the Phases

As noted, the phases in North American samples are associated regularly and robustly with deficits or deficiencies. Thus persons in phase I have higher job satisfaction than those in II, persons in phase II report higher levels than those assigned to III, and so on.

The transnational concurrent validity of the phases largely remains a challenge for the future, and hence the present tentativeness about this model-specific perspective. A growing collection of Canadian studies (e.g., Burke & Deszca, 1986) begins a replication of the U.S. results, which are substantially consistent (e.g., Golembiewski & Munzenrider, 1988). One study of a small Chinese sample also begins to replicate U.S. and Canadian results (Rowney & Cahoon, 1990).

It is also relevant that not only is an extensive replication underway in Japan, but provisional results for a cohort of $N = 387$ (Golembiewski et al., 1992) show much the same pattern of association of covariates with the phases in Japan as in U.S. and Canadian samples. Six marker variables are employed, which, avoiding details, deal with health status, job involvement, helplessness, satisfaction with work, job tension, and self-rated productivity. The Japanese sample permits 168 total paired comparisons of the six variables on the eight phases. In summary, the findings are as follows:

1. All six marker variables arrayed by the phases attain $p < .0001$ by ANOVA, with eta^2 approximating .136.
2. 81.5% of all paired comparisons are in the expected direction (e.g., health status is better for I's than II's, etc.).

3. 10.7% of all paired comparisons are statistically significant as well as in the expected direction, by least significant difference test, as modified for unequal subsample sizes.
4. 18.5% of all paired comparisons fall in a contrary direction.
5. 1.2% of all paired comparisons attain statistical significance and also fall in a contrary direction.

This pattern of covariation is the same as in North American samples for the same or similar marker variables, but the magnitude is not as great. For example, the latter samples often explain 15–20% of the variance as estimated by η^2 and have 90% or more of their paired comparisons falling in the expected direction, with 20–30% and usually more of the cases also attaining statistical significance.

The difference in magnitude and similarity in pattern of covariation can have several explanations. Thus the sample $N = 387$ is a composite from several work settings with a range of roles and national locations (Golembiewski et al., 1992), and this may dilute the strength of the associations while retaining their basic pattern.

Those results also may reflect cultural differences between Japanese and North American contexts. Or, more prosaically, the explanation may lie in the fact that, despite 387 cases, three of the phases still contain fewer than 10 cases each. These may induce heterogeneous variance that helps account for the lesser magnitudes of the associations sketched above with marker variables in Japan, as compared with much larger U.S. and Canadian samples.

Larger Japanese samples will provide useful perspective on this central particular. We do know that the bulk of the cases falling in an unexpected direction involve those three underpopulated phases in the Japanese replication.

DISCUSSION

Although far from decisive, both the generic and phase-specific perspectives support the cross-national (and perhaps cross-cultural) character of burnout. This constitutes a major conclusion, if clearly a preliminary one.

Viewed generically, burnout cannot be dismissed easily as somehow culturally restricted. That is, the concept has experienced a rapid diffusion over much of the world in a short period of time. Moreover, cross-national comparisons of responses to the MBI imply substantial generalizability of burnout subdomains. In addition, the findings from cross-national research—although still in the early stages—fit comfortably with the work on stress and burnout in the United States and Canada.

Considering only the phase model, research on burnout also seems to be broadly relevant. Thus the cross-national distributions of the phases suggest a major social problem—apparently less in some national settings and greater in others, but *always* substantial. Moreover, the literature lacks a reliable and valid sense of who has which degrees of burnout. The phase model provides just such estimates and, given the need for further replication, available research permits substantial optimism. Emerging research on cross-national replications of the associations of the phases with the same or similar marker variables, as in the Japanese example above, reinforces the model's relevance.

The analysis above also suggests eight positive thrusts for future research. First, burnout is no conceptual phantom. "It" seems a robust dimension of reality, in fact. The issues now involve moving on to more comprehensive kinds and levels of analysis.

Second, the results support additional cross-national and cross-cultural work. Available studies reveal patterns of burnout covariates much like those isolated in U.S. and Canadian settings. Specifically, the three subdomains of burnout seem to be perceived similarly by samples from different nations, but the levels of burnout seem to differ as do the phase distributions.

Third, the results above imply the value of more research using the three Maslach MBI subdomains for estimating burnout. In U.S. research, the MBI has come to dominate (Kilpatrick, 1986), and several lines of evidence permit a strong claim that its three subdomains tap fundamental dimensions of reality.

Fourth, although it may appear self-serving, results with the phase model pose theoretical and practical issues of great significance in cross-national or cross-cultural research. Consider the huge proportions in phases VI–VIII and the rare I's in available Japanese samples. On the face of it, these distributions are consistent with the common wisdom about major cultural features of Japanese life and imply the urgent need for remedial action. Although work is still in process, for example, we see the almost total absence of Japanese in phase I as reflecting their apparently great tendency to depersonalize—to repress feelings, to seek to "save face" for self and others, and so on (Boudreau & Golembiewski, 1989, 1990). Directly, Canadian and U.S. convenience populations approximate 5 to 10 times more I's than do three Japanese samples. For conceptual reasons beyond the present discussion but conveniently available elsewhere (e.g., Golembiewski & Munzenrider, 1988, pp. 24–27), high levels of depersonalization in the phase model's view of chronic onset will tend to trigger quickly in individuals a diminishing sense of personal accomplishment and then a heightened emotional exhaustion. That is to say, pervasive depersonalization implies numerous assignments to advanced phases. The tendency would be exacerbated by long work weeks and intense effort, both said to characterize many Japanese working lives.

Other explanations for the Japanese distributions of phases also require analysis. Thus, Burke (1989b) urged comparative tests of several measurement conventions underlying the phase model, e.g., are local medians better cutting points for the high/low distinctions on which the phase model rests, or is it more appropriate to rely on the "universal norms" that underlie the research on the phase model reviewed above? The latter norms are based on US 6, and those norms may be inappropriate for the Japanese cultural context. In either case, the phase model poses key questions for research and application.

Fifth, the dominant designs in available burnout research fall far short of ideals. For example, with few exceptions (e.g., Himle et al., 1986), burnout research relies on convenience populations. This leaves findings open to the charge that they are not generalizable. Large, stratified, random samples seem indicated.

Sixth, as in most behavioral research, what may be called "comparative operational analysis" has received little attention. Except for the MBI, the overlap of various burnout-related operational definitions remains in doubt and this in turn implies major imponderables in the interpretion of research.

Seventh, attempts to ameliorate burnout are rare (e.g., Golembiewski et al., 1987), and this fact cannot be tolerated. Note especially that, given the indication that stress and mental ill health are particularly virulent among employees in the developing and recently industrialized countries, remediation involving non-Western settings has a high priority. Relatedly, gender and cultural differences in the use and perceived effectiveness of coping strategies also merit attention.

Eighth, the burnout literature lacks an appropriate emphasis on objective or "hard" covariates. Self-reports dominate, but multiple measures, both hard and soft, have much to recommend them.

14

BURNOUT AS A DEVELOPMENTAL PROCESS: CONSIDERATION OF MODELS

Michael P. Leiter
Acadia University and Dalhousie University

Psychological burnout is a complex phenomenon, and research on this construct has reflected its complexity. Maslach (Chapter 2) emphasized the origins of burnout research in social, rather than clinical, psychology. However, this field of inquiry is characterized by a persistent tension between social and clinical perspectives. The term used to describe this research field, *burnout*, is the label for an extreme end state that shares many features with the clinical syndrome of depression (Firth, McIntee, McKeown, & Britton, 1986; Meier, 1984). Although the term appears to be an appropriate description for a distressing state experienced by human services workers overwhelmed by the emotional demands of their work, the activity occurring under the general label of burnout research has been concerned with much more than this extreme state. In fact, it is the contention of this chapter that the viability of burnout as a research area depends on a thorough understanding of the full range of experiences assessed in burnout studies. This goal will be facilitated by the development of models that articulate the relationships among components of burnout and their relationships with organizational and personal conditions.

Implied in this proposition is an interplay between conceptual and measurement issues. The phenomenological development of the primary measure of burnout, the Maslach Burnout Inventory (MBI) (Maslach & Jackson, 1981a, 1986), focused on providing an accurate description of career crises experienced by human service workers. Although the name of the scale adopts the colloquial term for an extreme crisis of this sort (see Meier, 1984; Pick & Leiter, 1991), the middle ranges of the measure appear to measure career crises that are less all-encompassing. The consideration of moderate ranges on the burnout measure leads to a consideration of scores on the opposite end of the MBI from burnout. It is argued that they measure a subjective state of professional effectiveness, not merely the absence of a crisis.

These considerations have implications for one's conceptualization of burnout and for developmental models of the syndrome. An emphasis on burnout as an extreme, almost clinical, condition is most compatible with a unidimensional measure of the construct rather than the more cumbersome and complex MBI. Again, developmental models focusing on the extreme state of burnout will tend to deemphasize moderate scores as stages along the way rather than states of inherent interest. In contrast, an approach viewing the MBI as measuring a wide range of experiences of efficacy and involvement in personal relationships would give equal emphasis to the full range of the scale.

These issues are discussed below in consideration of two models of burnout and a review of some of the research associated with these models. The implications of these considerations for burnout research are then discussed.

BURNOUT AS A UNITARY VERSUS A THREE-COMPONENT CONSTRUCT

The Three-Component MBI

From their interviews with human service providers, Maslach and Jackson (1981a, 1986) developed a 47-item scale that they reduced to the current 22-item scale on the basis of factor analyses of the responses of over 1,000 people in a series of studies. These studies resulted in a three-factor structure for the MBI, which has been confirmed with minor caveats by subsequent researchers (Evans & Fischer, 1989; Lee & Ashforth, 1990; Leiter, 1988b). The instructions Maslach and Jackson (1981a, 1986) present for scoring the MBI yield three scores, as opposed to the one-dimensional Tedium Measure or Burnout Measure of Pines, Aronson, & Kafry (1981). Maslach and Jackson have never presented instructions for calculating a unidimensional burnout score, but have stated that "a high degree of burnout is reflected in high scores on the Emotional Exhaustion and Depersonalization subscales and in low scores on the

Personal Accomplishment subscale" (Maslach & Jackson, 1986, p. 2). They go on to say that "the scores for each subscale are considered separately and are not combined into a single, total score" (p. 2).

The absence of a procedure for computing a unidimensional burnout score does not reflect a lack of quantitative skill on the part of the scale's developers or of a lack of demand for a unitary burnout score. It reflects considerations about the nature of burnout and is consistent with the developers' purpose of providing an empirically based burnout measure. Maslach and Jackson (1986) attributed their reluctance to combine the subscales to insufficient information regarding the nature of the three components and their interrelationships. As more is learned about the phenomenon of burnout, the complexity of the three-component definition of burnout has provided a conceptual richness that more than justifies the problems people encounter in thinking about the construct. It is my view that to some extent the vitality of the research activity regarding burnout is due to the complexity of the MBI. A unidimensional burnout measure would provide convenience at the cost of conceptual accuracy (cf. Chapter 2).

The Phase Model of Golembiewski and Munzenrider

Golembiewski and Munzenrider (1988) developed a phase model intended to reduce the MBI to a unitary measure (see also Chapter 13). They cite two advantages of their phase model. First, the phase model provides "a way to measure burnout in large aggregates, as well as to classify individuals in terms of the virulence of their particular cases" (p. 29). Second, they claim that the phase model is particularly suited to tracking the impact of interventions to the extent that "no other way of measuring burnout allows for [the phase model's degree of] flexibility or has such potential use for both research and practical interventions" (p. 30). These two claims are considered below after a description of the procedure for deriving the eight unidimensional phases from the three-part MBI.

The definition of phases

The phase model of Golembiewski and Munzenrider (1988) is derived from a version of the MBI that they modified to increase its appropriateness to people working outside of human service professions on which Maslach and Jackson (1981a) originally developed the scale. Their modifications include changes in the rating instructions: rather than the current MBI frequency scale ranging from 0 "never" to 6 "every day," items are rated from 1 "very much UNLIKE me" to 7 "very much LIKE me." Golembiewski and Munzenrider drop two MBI items entirely and add three new ones. They claim that the resulting measure produces three factors clustering into emotional exhaustion, deperson-alization, and personal accomplishment, with a few items showing cross-loadings between two factors. Golembiewski and Munzenrider report a factor analysis supporting this three-factor structure, but this report provides insufficient information regarding the method of factor analysis used, and the authors' rationale and criteria for defining a factor. The extent to which data from workers outside human service professions results in a three-factor structure is of particular interest in light of recent analyses of the MBI (Evans & Fischer, 1989).

Another significant change from the MBI is that the Golembiewski and Munzenrider items refer to coworkers rather than to service recipients. This change permits the scale to be relevant to people outside of human service professions. The extent to which it indicates a measure of something distinctly different from burnout in the sense defined by Maslach and Jackson (1981a) and Cherniss (1980a) is unclear (see Garden, 1987b). The definitions of these authors were consistent with the work of Freudenberger (1978), who approached burnout as a specific occupational problem of human service providers, in which the demands of the personal supportive relationship between service provider and recipient played a central role. Further, the value system of the organizational culture of public service agencies played an important role in these conceptualizations of burnout (Cherniss, 1989b; Leiter, 1991b).

In any case, their modified MBI produces three scores, labeled identically to those from the original MBI: emotional exhaustion, depersonalization, and personal accomplishment. Each of the three scores is compared to norms from a large normative sample and assigned a binary value of high or low. The eight phases represent the 23 ways of combining three binary values. These are arranged in a sequence that Golembiewski and Munzenrider refer to as progressive "virulence," although they neglect to provide a convincing theoretical rationale for this attribution of virulence. The end points of this progression are rather straightforward: the state of scoring in the burnout direction on all three subscales is the most virulent; the state of scoring in the opposite direction on all three subscales is the least virulent. The intervening phases are defined by alternating depersonalization each phase; alternating personal accomplishment every two phases; and assigning low emotional exhaustion to the first four phases and high emotional exhaustion to the second four (Table 14-1).

Limitations of the phase model

In Leiter (1989), I criticized this model as reducing burnout to the single dimension of emotional exhaustion. This dimension of burnout has consistently been found to be more strongly correlated with environmental stressors (see Chapter 11). Whereas the three aspects of burnout are correlated with one another, emotional exhaustion scores show a near-linear relationship with the phases. Therefore, any measure of environmental demands or hassles that are correlated with emotional exhaustion will be correlated with the burnout phases. Consequently, the relationships of outcome measures or environmental condi-

TABLE 14-1   Phases of Burnout

Phase	Depersonalization	Personal accomplishment (reversed)	Emotional exhaustion
I	Low	Low	Low
II	High	Low	Low
III	Low	High	Low
IV	High	High	Low
V	Low	Low	High
VI	High	Low	High
VII	Low	High	High
VIII	High	High	High

Note: After Golembiewski & Munzenrider, 1988, p. 28.

tions with these phases do *not* reflect anything about progressive virulence of burnout that was not apparent from the correlations of such measures with emotional exhaustion.

Table 14-2 displays the mean values of emotional exhaustion across the eight phases for a sample of people working outside of human service professions, based on the original MBI. Divisions into high and low were based on population norms reported in Maslach and Jackson (1986). As such, these data are not presented as a direct investigation of the Golembiewski and Munzenrider measure. However, to the extent that the Golembiewski and Munzenrider measure is consistent with the original MBI—as emphasized by Golembiewski and Munzenrider—this analysis should have relevance to their measure. As seen in Table 14-2, the progression of phases is highly consistent with increases in emotional exhaustion. The two exceptions to a consistent progression are

TABLE 14-2   Values of Emotional Exhaustion across Burnout Phases

Phase	N	Mean	SD	95% confidence interval
I	97	8.92	5.63	7.79–10.06
II	20	13.85	4.12	11.92–15.77
III	123	9.69	5.23	8.75–10.62
IV	41	13.43	4.67	11.96–14.91
V	19	27.78	4.61	25.56–30.01
VI	26	31.00	6.96	28.18–33.81
VII	41	29.07	5.55	27.32–30.82
VIII	90	32.03	7.69	30.42–33.63
Total	457	18.15	11.58	17.07–19.22

Note: After Clark, 1990.

phase II and phase VI, which are higher than expected. These are relatively rare phases, including only 10% of the sample. Thus, they would have a negligible influence on the overall relationship between exhaustion and phases. The correlation between phase number and emotional exhaustion was .81, in contrast with .62 with depersonalization and −.42 with personal accomplishment. In summary, these considerations indicate a high degree of redundancy between the emotional exhaustion aspect of burnout and the phase model. It correspondingly reduces the role of depersonalization and personal accomplishment in the model of burnout.

Classification

The phase model does help in classification by reducing the combinations of the three burnout subscales to eight phases. The ranges of potential scores on the three original MBI subscales are emotional exhaustion, 0–54; depersonalization, 0–30; personal accomplishment, 0–48. These ranges produce 83,545 different scores ($55 \times 31 \times 49$). The phase model reduces this combination on the order of 10,000! (The number of combinations one would encounter in actual practice would be smaller due to the correlations among the MBI subscales and their restricted range.)

To Golembiewski and Munzenrider, the phase model assists categorization not only by reducing the number of combinations to eight, but also by producing categories that constitute an ordinal scale in which the burnout phases are arranged in order of progressive virulence. A consistent theme in their research has been establishing that people scoring at a given phase are experiencing more distress than people at a lower phase and less distress than those at a higher phase.

Suitability to large aggregates

Golembiewski and Munzenrider do not share their reasoning behind their claim that the phase model is particularly suitable to large aggregates. On the one hand, the phase model allows one to state that there are X people in phase Y, which may be more palatable to some audiences than information about means, medians, and standard deviations. On the other hand, categorization gives credence to boundaries that are entirely mathematical constructs. The assignment of people into phases increases the potential for assigning labels with potentially negative impact on self-esteem or a person's employment status. It may also increase the inclination to view burnout as a psychiatric syndrome rather than as a social construct, in that the labeling process emphasizes burnout as a feature of an individual.

Benefits: sensitivity to interventions

Golembiewski and Munzenrider do not share their reasoning behind their claim that the phase model is particularly sensitive to the impact of interventions. In fact the reduction of the complex MBI subscale scores to eight categories could only reduce the sensitivity of the scale to changes. As I indicated in Leiter (1989), the phase model only registers changes that cross the mean of an MBI scale. It is insensitive to large changes that do not cross the mean. For example, a person whose score changed from 1 point above the mean on the emotional exhaustion subscale to 2 SD above the mean during an intervention study would register no difference in burnout phase. Another person whose score changed from 1 point above the mean to 1 point below the mean on emotional exhaustion while remaining steady on the other subscales would register a change of four phases! This is not so much a matter of enhanced sensitivity as enhanced emphasis of small changes around the mean of the three subscales. Someone using the full emotional exhaustion subscale would register small changes near the mean with equal sensitivity but would consider them relatively minor in comparison to changes of a standard deviation or two, regardless of whether these changes crossed the mean. The phase model would consider a change of a point or two across the mean of equal weight as a large change across the mean and of much greater importance than a large change on one side of the mean. There may be a good reason for this emphasis, but it is not readily apparent.

Relative benefits of scoring approaches

In light of the above discussion, it appears that the strongest argument in favor of the phase model is categorization: it does reduce a complex three-part measure to eight phases. It remains to be seen whether eight is the ideal number of phases or whether the method Golembiewski and Munzenrider use to calculate the phases is best suited to the purposes of those who use the instrument. The first issue is largely one of sensitivity; the second is one of substance. Both are relevant considerations in one's choice in approach to a measure.

This discussion of the sensitivity of the MBI concerns the extent to which different scores on the instrument reflect valid differences in the state of the respondents. Although it is unlikely that the MBI actually differentiates among 80,000 psychological states, it may make meaningful distinctions among more than eight. Maslach and Jackson (1986) encourage practitioners to differentiate among high, medium, and low scoring respondents on the three subscales when providing feedback to respondents. Numerical feedback regarding a psychometric measure is rarely appropriate; its interpretation requires a context of descriptive statistics that are often unavailable to the general public. It also requires a much greater degree of explanation from the researcher to make the feedback intelligible. A label of high or low or a phase number is a direct way of providing general feedback that provides pertinent information.

The introduction of arbitrary boundaries and the reduction to a unidimensional scale are less appropriate to most research agendas. The extensive range of correlational studies using the MBI have found burnout to be linearly related to a wide variety of organizational conditions and outcomes (see

Kleiber & Enzmann, 1990). The median split procedure obscures the range and nature of these linear relationships by reducing each dimension to a binary scale. The domination of the phase model by the emotional exhaustion dimension deemphasizes the contribution of depersonalization and personal accomplishment to the burnout syndrome, thereby making burnout more similar to general occupational stress. Handy (1988) pointed out that much of the research on burnout has followed trails already explored in research on occupational stress. She argued that the viability of burnout as a research area depended on its generating new perspectives. It is the focus in burnout research on personal relationships and on professional efficacy that holds the greatest potential for a distinct contribution to organizational and social psychology (see Chapters 4 and 8).

BURNOUT AS A DEVELOPMENTAL PROCESS

Closely related to the issue of burnout dimensions is that of development. Many studies have attempted to identify conditions that bring about burnout, but it is the relationships among the three MBI subscales that have presented the more challenging problems. The central question here is whether there is a regular temporal sequence among the three MBI components, in that one aspect of burnout arises first, in response to environmental conditions, and in turn prompts the development of the other aspects of burnout. This view contrasts with one in which all three aspects of burnout arise relatively independently in response to similar, or related, environmental conditions. A few models of these relationships have been proposed, but there has not been conclusive research on the issues. These issues are discussed below.

Developmental Relationships among the MBI Subscales

My own view of the relationships among the MBI subscales has changed from a sequential developmental model to a mixed sequential and parallel development model. Originally, Leiter and Maslach (1988) proposed a sequence of relationships among the MBI subscales such that emotional exhaustion arises first in response to demanding environment, as human service workers are overwhelmed by the personal demands of a job. Increased exhaustion in turn brings about depersonalization as workers attempt to gain emotional distance from service recipients as a way of coping with the exhaustion. The exhaustion and the impoverished personal relationships with service recipients in turn diminish the workers' sense of personal accomplishment as the work loses its meaning.

More recent work has maintained the relationship between exhaustion and depersonalization described above, but has recast the relationship of personal accomplishment with the other two components of burnout. That is, the current model (Leiter, 1990b, 1991a) depicts depersonalization as a function of emotional exhaustion that mediates most of the impact of environmental conditions on depersonalization (see Figure 14-1). In contrast, the model in Figure 14-1 depicts personal accomplishment as a function of the work environment. In general, the model proposes that demanding aspects of the environment (e.g., workload, personal conflict, hassles) aggravate exhaustion, which in turn contributes to increased depersonalization, while the presence of resources (social support, opportunities for skill enhancement) influences personal accomplishment. For the most part, these two aspects of burnout have distinct predictors, although some conditions, such as coping styles, appear to contribute to both exhaustion and diminished accomplishment. Further, demands and resources are not entirely independent: in light of the somewhat circular definition of resources and demands in stress literature (Hobfoll, 1989; Lazarus & Folkman, 1984), it is not surprising that work environments that participants perceive to be overly demanding are also rated as offering insufficient resources.

This model proposes that the diminished personal accomplishment component of burnout develops in parallel with the emotional exhaustion component,

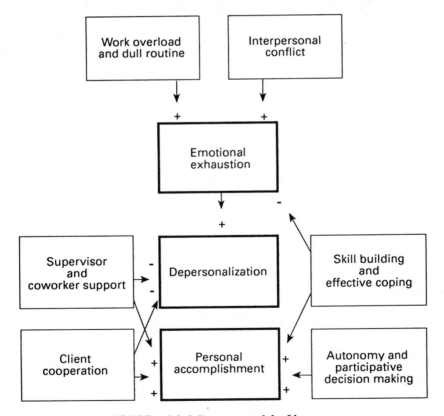

FIGURE   14-1 Process model of burnout.

as they arise as reactions to different aspects of work environments that pose difficulties for human service workers. This view is in contrast to the previous model (Leiter & Maslach, 1988), which proposed an entirely internal model in which emotional exhaustion triggered the whole burnout syndrome in response to environmental stressors. To a considerable and meaningful extent, the relationship between these two components of burnout exists external to individuals. It exists within their social context, in the provision of organizational resources, in the personal conflict among colleagues and service recipients, and in the pressure of emotional demands. From this perspective, the mild but persistent correlation between emotional exhaustion and personal accomplishment (Maslach & Jackson, 1986) arises because of consistent relationships among aspects of work environments: work settings characterized by excessive workloads and personal conflict are often experienced by their participants as lacking in organizational supports necessary for effective professional functioning (Cherniss, 1980a; Hackman, 1986). This is not to say that inadequate support is simply another way of phrasing excessive demand, but that work settings that persistently deplete the emotional energy of its personnel are likely to provide inadequate resources in terms of social support or opportunities for professional development.

In summary, the model developed by Leiter and Maslach (1988) through Leiter (1991a) is a process model of burnout in which time continues to be an important dimension. From this perspective, people do not suddenly become burned out, nor are high levels of exhaustion and a diminished sense of accomplishment invariable aspects of some people. Instead, people move toward increased professional efficacy or toward burnout as a function of their personal reaction to persistent aspects of their work environment. The relationships found among the three aspects of burnout reflect processes that are occurring within the person and consistencies within the organizational context. Although the model was originally developed on cross-sectional data, it has been successfully tested with longitudinal data (Lee & Ashforth, in press; Leiter, 1990b). It is consistent with other research (Gaines & Jermier, 1983; Lee & Ashforth, 1990; Shirom, 1989) in giving a central role to the emotional exhaustion aspect of the syndrome.

Development in the Phase Model

Golembiewski and Munzenrider acknowledged that their phase model is not meant to reflect a strictly temporal development in burnout, although they do project temporal progress through so-called flight paths that individuals may follow during the development of burnout. They distinguish phase sequences for acute and chronic onset of burnout. Their description of acute onset largely encompasses an acute stress reaction: a person reacts to a negative life event (e.g., loss of a spouse) with increased emotional exhaustion (phase V) followed by a simultaneous increase in depersonalization and a decrease in personal accomplishment (phase VIII). The processes through which acutely enhanced emotional exhaustion leads to changes in the other two components of burnout are not explained. This sequence is dominated by emotional exhaustion, which provides a good indication of psychological stress. Although this is a valid way to monitor stress reactions in occupational settings, it is difficult to determine if this sequence introduces anything distinctly characteristic of burnout.

Golembiewski and Munzenrider's end state of chronic burnout is closer to the way other researchers (Cherniss, 1980a; Maslach & Jackson, 1986) conceptualized the construct as a chronic condition but differs in the process through which the condition arises. Their chronic onset sequence begins with increased depersonalization (phase II) followed by an accompanied decrease in personal accomplishment (phase IV) ending with an accompanied increase in emotional exhaustion (phase VIII). From this perspective, a person experiencing burnout in the context of a difficult occupational and social context is indistinguishable from one who is reacting to stressful life events in terms of their burnout phase. The difference lies in the process through which they reached this end state.

Test of Developmental Sequences

Lee and Ashforth (in press) provided a direct contrast of the chronic sequence described for the phase model and the model postulated in Leiter and Maslach (1988). They analyzed data from human service workers (urban public welfare agency workers) using the standard MBI expanded by changing every item referring to recipients to two items: one referring to clients and an identical one referring to subordinates. Incidentally, they reported a .51 correlation between the paired subordinate/client items, suggesting that although they share common features, they are far from identical constructs. In any case, their expanded MBI provides something of a compromise between the standard MBI with which the Leiter model was produced, and Golembiewski and Munzenrider's revised MBI.

Lee and Ashforth's analysis of longitudinal data from two surveys separated by 8 months clearly supported the prediction of the Leiter and Maslach (1988) model that emotional exhaustion was a precursor of depersonalization. They did not fir. i the neat sequence predicted in that study of depersonalization subsequently diminishing personal accomplishment. Their derived model depicted personal accomplishment as a function of emotional exhaustion rather than as a parallel construct as in the Leiter (1990b, 1991a) model. However, they note that the relationships of personal accomplishment with the other aspects of burnout was unclear, and may be subject to influence from various aspects of a work setting.

Problems in Testing Developmental Sequences

The testing of developmental process models of burnout presents a variety of challenges in conceptualization, data collection, and analysis. One issue is the selection of an appropriate time lag between surveys. The subscales of the MBI are fairly stable measures. Lee and Ashforth (in press) report 8-month test–retest correlations for emotional exhaustion, depersonalization, and personal accomplishment of .74, .72, and .65, respectively. Other longitudinal studies (Golembiewski and Munzenrider, 1988; Jackson, Schwab, & Schuler, 1986; Leiter, 1990b) report similar degrees of consistency (see also Chapter 12). Testing a process model of burnout requires demonstration of consistent patterns of development against such a backdrop of consistency.

Longitudinal studies of burnout assume that developments occur over a matter of months. Jackson, Schwab, and Schuler (1986) used an interval of a year; Lee and Ashforth (in press) used 8 months; Leiter (1990b) used an interval of 6 months. These time intervals decrease the extent of interference involved in the repeated measures procedure of administering the identical questionnaire package. These intervals also imply a certain degree of resistance to change. An increase in one's level of depersonalization is not expected to immediately follow enhanced emotional exhaustion among human service workers, but rather to follow prolonged periods of exhaustion accompanied by a history of unsuccessful attempts to address the sources of exhaustion. Whereas depersonalization runs contrary to the ethical principles of human service professions and widely shared assumptions regarding the efficacy of close personal relationships in enhancing therapeutic relationships, it is not readily accepted as a response to emotional depletion. However, persistent conditions of exhaustion in a situation offering few avenues of effective change are expected to wear down that resistance in a matter of months. An excessively short interval between surveys may miss delayed changes in depersonalization.

An overly lengthy interval creates various problems. Longer intervals increase the probability of subject mortality as people change jobs. Even if they remain in the same organization, their role may change in a way that is pertinent to the measures. Longer intervals increase the probability that major organizational initiatives or crises may occur. Also, longer intervals permit the unfolding of more complex processes. Events may occur that enhance emotional exhaustion, followed by subsequent events that diminish it, thereby obscuring whatever relationship it maintains with depersonalization.

A process model of burnout includes relationships between environmental conditions and burnout as well as relationships among the three components of the syndrome. Each of these relationships may occur at a different pace. An increase of workload or interpersonal conflict may affect workers immediately, while an enhancement of coworker relationships or skill utilization may take longer to make an impact on exhaustion or personal accomplishment. Daily stress logs may be more appropriate for examining some aspects of the burnout process, but widely spaced questionnaire surveys may be more appropriate for others. It is likely that a thorough examination of a process model of burnout will involve the integration of a variety of research methods. As there is no theoretically correct developmental pace for the development of burnout, the choice of time lag remains a question that has to be empirical.

CONCLUSIONS AND RECOMMENDATIONS

This chapter raised two concerns regarding the definition of burnout. First, is burnout a unitary construct? Can people be categorized as more or less burned out on a linear scale, or is burnout a more complex syndrome? Second, to what extent is temporal development intrinsic to the concept of burnout? Are there systematic causal relationships among the components of burnout as measured by the MBI?

The three-dimensional quality of burnout increases its potential to make a distinct contribution to organizational psychology. Although the central dimension of emotional exhaustion tends to dominate the relationships of burnout with environmental conditions, it undermines the distinctiveness of the syndrome. An exclusive concern with the exhaustion dimension lessens the contrast between burnout research and work on general occupational stress, as well as burnout's particular relevance to human service professions. The dimensions of depersonalization and personal accomplishment provide a direct focus on two issues of central concern to human service providers: therapeutic relationships with service recipients and the development of professional efficacy.

The use of a tripartite definition of burnout requires an articulation of the relationships among the components. Although it is reasonable to propose that depersonalization represents an attempt to cope with the experience of emotional exhaustion by gaining psychological distance from service recipients, this functional relationship has not been demonstrated. Conceptualizing personal accomplishment as an indicator of professional efficacy brings this dimension of burnout within a coping framework as well. These ideas suggest that a fully articulated model of burnout development will include a complex process of appraisal and coping with the emotional demands of human service occupations.

The experience of burnout is a complex phenomenon. Attempts to understand its relationships with stressful and supportive aspects of work environments require the use of analytical procedures that can address that complexity. The examination of correlations or the reporting of redundant comparisons between burnout phases leaves too many important questions unanswered. Model testing approaches, such as LISREL (Jöreskog & Sörbom, 1989), provide a means of examining the relationships of the MBI subscales with environmental or personal variables while simultaneously addressing the relationships among the three aspects of burnout. In particular, the articulation and contrast of burnout theories, such as that performed by Lee and Ashforth (in press), are essential to the development of a coherent theory.

The use of model testing analyses expands the scope of the concepts one can explore with quantitative data. The field is in a process of complementary development in which clarification of theory interacts with advances in analytical approaches. However, there continue to be limitations. Little can be gained toward understanding a developmental process through examining samples comprising a large proportion of individuals who have attained equilibrium. Testing of theories of the developmental processes in burnout requires more research with populations that are undergoing major transitions in their professional development, such as students graduating from professional degree programs. Another population of interest is people working in public social service agencies that are confronting major reductions of funding. Financial restraint programs of governments in North America and the United Kingdom have led to the constraint of many social programs. It is likely that the increase in workload and reduction of resources associated with these cutbacks may have a direct impact on the incidence of burnout (Leiter, 1990a).

In light of the central role given to developmental concerns in burnout research, there is a considerable contribution to be gained from studies with repeated measures designs. However, it is unlikely that longitudinal questionnaire studies will on their own provide conclusive insights into the development of the syndrome. There is a major role to be played by observational studies, in-depth interviewing, and studies from a grounded theory development perspective. The concept of professional efficacy has the potential to make a contribution to the field, but considerable work remains to define the limits to the applicability of the concept. The wealth of material in this book and in the conference that inspired it is a clear sign that there is a strong interest in pursuing these goals.

As a final note, there is a strong argument for changing the name of this research field. The term *burnout* is most relevant to one end of a continuum, the other end of which is characterized by effective professional involvement (Leiter, 1992). To direct the focus onto this full range, and to consider the more limited states measured in the middle ranges of the MBI, the label of "professional efficacy" or "career crisis" would be more informative and would help to generate new ideas.

V

OUTLOOK

15

THE FUTURE OF BURNOUT

Wilmar B. Schaufeli
University of Nijmegen

Christina Maslach
University of California, Berkeley

Tadeusz Marek
Jagiellonian University

In this closing section of the book we will try to strike a balance and see what the future holds for the burnout construct. First: What do we have? What has been achieved by the recent developments in theory and research? What contributions have been made by the chapters included in this book? And what issues are still unresolved? Of course, we can only deal with these issues in a somewhat global manner. The reader is referred to the chapters for more detailed information.

Second: What do we need? What still remains to be done in the field of burnout? We will elaborate on six directions for future theorizing and research. In our opinion, the current volume provides a number of suggestions that could be used as guidelines.

WHAT DO WE HAVE?

A Relevant Construct

In burnout we have a concept that is highly relevant, socially as well as academically. As was described in the first chapter, burnout emerged as a social problem rather than as a scholarly issue. As far as we can see, it will continue to be a relevant social topic in the near future as well because the sociocultural and socioeconomic developments that foster burnout remain active. For instance, the increasing individualization in our society affects the traditional social structure (e.g., community, family). This causes people to rely more and more on professionals for help, advice, support, and care. Accordingly, the workload in the human services increases, but at the same time there is considerable political and economic pressure to cut back public funds, which leads to an understaffing of these institutions. In other words, more work has to be performed by fewer people. It is reasonable to assume that this situation contributes to the emotional depletion of the professionals involved.

During the last decade burnout has sparked a great deal of theoretical thinking and empirical research. This book demonstrates that burnout has been placed successfully on the academic agenda. It seems to be a concept that can guide serious theoretical as well as empirical investigation. Moreover, the burnout concept appears to be a fruitful extension and deepening of the traditional occupational stress research area. There is a great need to study particular work-related strains in particular work environments and this is precisely the type of contribution that burnout research tries to make.

A Construct To Be Conceptually Integrated

With burnout we have a construct with the potential to be integrated in a wider theoretical framework. The contributions in this book cover three levels of conceptual integration. First, burnout is considered from four general psychological perspectives. Buunk and Schaufeli (Chapter 4) investigate burnout from a social comparison point of view, drawing on the seminal work of Schachter, among others, on social affiliation. Hobfoll and Freedy (Chapter 7) apply a general stress theory to burnout—the recently developed conservation of resources approach. Cherniss (Chapter 8) uses the early work of Hall and Bandura on motivation to develop the concept of professional self-efficacy, which he uses as a unifying theoretical construct to understand burnout. Finally, Cox, Kuk, and Leiter (Chapter 11) try to integrate burnout conceptually within the transactional model of occupational stress, which was developed by the first author.

Second, in a number of contributions, burnout is linked somewhat more eclectically with elements from several theoretical approaches. For instance, in order to understand burnout, Pines (Chapter 3), Burisch (Chapter 5), Hallsten (Chapter 6), Winnubst (Chapter 9), and Noworol et al. (Chapter 10) borrow concepts from existential psychology, action theory, ego psychology, organizational theory, and the psychology of creativity, respectively.

Finally, several authors offer a description of a model of burnout that is based mainly on the outcomes of their empirical research, i.e., Maslach's multidimensional perspective (Chapter 2), Golembiewski's phase model (Chapter 13), and Leiter's process model (Chapter 14).

A Common Language

With burnout we have a common language to denote a particular phenomenon. The definition of burnout has been an issue from the very beginning, as indicated in the first chapter. In recent years, there has been more consensual agreement on an operational definition of burnout, largely because of the development of validated research measures, particularly the Maslach Burnout Inventory (MBI). Consequently, researchers now have a common language for studying burnout and can make direct comparisons between their own findings and those of others—thus allowing new studies to build on the contributions of previous ones. As shown by Schaufeli, Enzmann, and Girault (Chapter 12), the MBI has encouraging psychometric properties. Moreover, Golembiewski et al. demonstrated in Chapter 13 that this instrument can be used in cross-national studies. However, there is still discussion and debate about the definition of burnout, as indicated by the contributions of Burisch (Chapter 5) and Hallsten (Chapter 6).

A Common Occupational Context

With burnout we have a common occupational context. That is, burnout has been studied primarily in the human services where professionals are doing "people" work of some kind. What is special about their work is that their tools are their own social skills, attitudes, and personality characteristics in addition to their professional technical abilities. The professionals' relationship with the recipient is the vehicle for change and hence the source of accomplishment (or of failure). At the same time, this relationship is demanding and exhausting by its very nature—recipients are troubled or suffering, and they are asking for assistance of some kind. Thus, in contrast to other types of occupational strain, burnout results from interpersonal processes with recipients as well as from an interplay of individual and organizational factors. Viewed from this perspective, burnout has a specific etiology that is linked to a particular professional domain.

Although burnout is studied most intensively in human service professions, similar processes might occur in other areas as well. Some authors, such as

Pines (Chapter 3), Burisch (Chapter 5), Hallsten (Chapter 6), and Winnubst (Chapter 9), argue that burnout is a much more general phenomenon that is not restricted to professionals who work with other people in some capacity.

Specific Methodologies

With burnout we have some specific methodologies that can be applied in research. For instance, the "phase model" of Golembiewski, Scherb, and Boudreau (Chapter 13) classifies individuals in eight different categories of burnout. Although this model has been criticized by Leiter (Chapter 14) on several grounds, the authors can claim a cross-national validity for their model.

WHAT DO WE NEED?

Theory-Driven Research

We need more theory-driven studies on burnout. This book shows that there is an alternative to the blind empiricism that was typical for so many early studies on burnout (see Chapter 1). As indicated above, we now have a number of interesting approaches that try to integrate burnout into larger conceptual frameworks such as social comparison theory (Chapter 4), general stress theory (Chapter 7), occupational stress theory (Chapter 11), and motivational theory (Chapter 8). These recent developments open the possibility of studying burnout within a broader theoretical context: i.e., more general psychological concepts can be applied to burnout. The main focus of this type of study is the theoretical integration of the burnout construct into larger, overarching psychological frameworks.

A second avenue of theory-driven research is more problem-oriented and involves the etiology of burnout. In this book, three types of etiological factors of burnout are distinguished: interpersonal, individual, and organizational. In each of the corresponding three sections of the book, theoretical notions and models are introduced that can function as cornerstones for empirical research on the etiology of burnout. Although some of the proposed models are not very easy to test empirically, e.g., Burisch' action model, they do provide some guidelines for theory-driven research.

Finally, theory-driven research (including etiological studies) should concentrate on the process on burnout rather than on the end state. This calls not only for longitudinal research designs but for particular conceptualizations, e.g., hypotheses about the sequential relationships of the burnout dimensions. Many approaches to burnout in this book explicitly stress the importance of such process models of burnout (Chapters 2, 5, 6, 14).

Different Assessment Methods

We need to expand the methodologies used and include measures other than self-report, such as indices of job performance, turnover and absenteeism rates, ratings by others (clients, supervisors, peers, family), physiological assessments of health, and so forth. The self-report indicators of burnout that have been employed until now lack any validation against such kinds of measures (Chapter 12).

Base Rate Information

We need more information about burnout base rates. There are no descriptive statistics that will allow us to answer such basic questions as the following: What is the frequency of burnout? Do many people experience it or just a few? Is the burnout rate 1%, 10%, 50%, 90%? Does the burnout rate vary by occupation? Once someone experiences burnout, how long does it last? Do people have repeated episodes of burnout, or does it only happen once? Do most people recover from burnout, or is their subsequent work impaired in some way? Although there has been much speculation about the answers to these questions, and various percentage rates have been proposed, there is as yet no solid set of statistics on which these answers can be based. Even when large surveys have been done, which could possibly generate some base rate statistics, they have often been limited by having self-selected, nonrepresentative subject populations (Chapter 12).

Cross-National Research

We need more cross-national research. Now that burnout research has spread outside the English-speaking countries, the need for systematic cross-national research is more urgent. Most studies on burnout have been conducted in the United States, but do the conclusions from these studies hold in the other national or cultural settings? This is by no means self-evident because important differences exist between countries in value systems (e.g., the meaning of work), organization of services (e.g., state or private), and the type of systems for health care and social security (e.g., in some countries, such as The Netherlands, employees can be declared work-disabled on psychological grounds, whereas in other countries this is impossible).

It appears that the MBI can be used in cross-national settings (Chapter 13). Apparently, burnout is experienced similarly in different countries: i.e., identical dimensions were found across countries. However, the question of whether levels and correlates of burnout are also comparable across countries is still an open one.

Criterion Levels

We need criterion levels for burnout. That is, at what point or level of experience does burnout become a true problem? Is it really a serious enough problem to warrant serious attention? The position taken by some critics is that it is a common experience for people to dislike their job at some point but that this dislike (or "burnout") does not lead people to have any major difficulties in doing their work. In other words, it has not been shown conclusively that burnout is causally related to objective outcomes of consequence and importance. Thus, another task for researchers is to provide empirical evidence for the seriousness of burnout as a social problem.

There are three areas where this could be done: the job (with such outcomes as turnover, job withdrawal behaviors, deterioration of job performance), the home (marital or familial conflict), and the individual (decline in physical and mental health). This underlines the previous arguments for the inclusion of measures that are not exclusively based on self-reports.

When more is known about criterion levels of burnout, valid cutoff points (e.g., on the MBI) can be determined that would allow for a differentiation of "burnout cases" from "noncases." In other words, individual diagnosis of burnout will be possible when we have adequate information on criterion levels.

Longitudinal Research Designs

We need longitudinal research designs. It is only through longitudinal designs that causal relationships can be established between burnout and both its precipitating factors and its outcomes. Although discussions of burnout are often filled with statements of presumed cause and effect, they are often based on the results of correlational, cross-sectional designs that do not permit such causal inferences (Chapter 14). In addition to providing better evidence for causal relationships, longitudinal studies would yield valuable information about the development and successive phases of burnout. The latter was put forward previously when the need for understanding the process of burnout was discussed. Longitudinal studies are very costly in terms of time, effort, and resources, and thus present a large challenge to any researcher. However, it is clear that such studies will be critical to our understanding of burnout.

Evaluation Research

We need good evaluation research with respect to burnout interventions. Although there are many ideas about how to deal with burnout (Chapters 3, 7, 8, 11), and some of these ideas may have been put into operation, there is virtually no concrete empirical evidence as to whether any of these interventions have actually worked. Part of the problem here is that researchers are often not involved in the implementation of an intervention, and thus are not in a position to define the critical variables (e.g., what behavior is predicted to change) or to collect the necessary information for evaluation (e.g., baseline measures prior to the intervention).

Another issue is that the long-term impact of any intervention is not easy to assess, largely because of the problems involved in doing longitudinal research. However, if a major goal of studying burnout is to determine effective strategies for dealing with it, then evaluation research will be absolutely crucial.

CONCLUSION

And so, at last, we come to the end of our journey through the field of burnout. What should be evident from this trip is that there is now a more clearly demarcated map of the burnout terrain. We know more (although not all) about the phenomenon of burnout, its antecedents and its consequences, and we also know more about what needs to be discovered. Although we can recognize some of the important landmarks of burnout, our map would be improved if we could fill in many of the critical details. Our group of international travelers has spoken in many languages, both cultural and conceptual, and has added much to both our knowledge and skills for the next stage of the trip. We have made some great strides since the beginning of the work on burnout, but there are distant horizons that have yet to be reached. It is our hope that this volume has functioned as an essential guidebook for this journey, and that it has succeeded in pointing out future paths for the next decade of travelers along the burnout road.

REFERENCES*

Abramson, L. Y., Metalsky, G. I., & Alloy, L. B. (1989). Hopelessness depression: A theory-based subtype of depression. *Psychological Review, 96,* 358–372.

Abramson, L. Y., Seligman, M. E. P., & Teasdale, J. D. (1978). Learned helplessness in humans: Critique and reformulation. *Journal of Abnormal Psychology, 87,* 49–74.

Ackerley, G. D., Burnell, J., Holder, D. C., & Kurdek, L. A. (1988). Burnout among licensed psychologists. *Professional Psychology: Research and Practice, 19,* 624–631.

Adler, N. J., Doktor, R., & Redding, S. G. (1986). From the Atlantic to the Pacific century: Cross-cultural management reviewed. *Journal of Management, 12,* 295–318.

Affleck, G., & Tennen, H. (1991). Social comparison and coping with major medical problems. In J. Suls & T. A. Wills (Eds.), *Social comparison: Contemporary theory and research* (pp. 369–394). Hillsdale, NJ: Erlbaum.

Affleck, G., Tennen, H., Pfeiffer, C., Fiflied, J., & Rowe, J. (1987). Downward comparison and coping with serious medical problems. *American Journal of Orthopsychiatry, 57,* 570–578.

Ahmavaara, Y. (1954). Transformational analysis of factorial data. *Annals of the Academy of Science Fennicae,* Series B, *881,* 54–59.

*Throughout this section, asterisks indicate the "empirical and cross-national" citations referred to on p. 218.

Amabile, T. M. (1983). *The social psychology of creativity.* New York: Springer.

American Psychiatric Association. (1980). *Diagnostic and statistical manual of the mental disorders.* 3rd Ed. Washington, DC: Author.

Anderson, M. B. G., & Iwanicki, E. F. (1981). *The burnout syndrome and its relationship to teacher motivation.* Paper presented at the Annual Meeting of the American Educational Research Association, Los Angeles.

Antonovsky, A. (1987). *Unraveling the mystery of health: How people manage stress and stay well.* San Francisco: Jossey-Bass.

Appley, M. H., & Trumbull, R. (Eds.). (1986). *Dynamics of stress: Physiological, psychological and social perspectives.* New York: Plenum Press.

Argyris, C. (1964). *Integrating the individual and the organization.* New York: John Wiley and Sons.

Aronson, E., Pines, A., & Kafry, D. (1983). *Ausgebrannt. Vom Überdruss zur Selbstentfaltung* [Burnout. From tedium to personal growth]. Stuttgart: Klett-Cotta.

Aström, S., Nilsson, M., Norberg, A., Sandman, P-O., & Winblad, B. (1991). Staff burnout in dementia care: Relations to empathy and attitudes. *International Journal of Nursing Studies, 28,* 65–77.

Bandura, A. (1977). Self-efficacy: Toward a unifying theory of behavioral change. *Psychological Review, 84,* 191–215.

Bandura, A. (1982). Self-efficacy mechanism in human agency. *American Psychologist, 37,* 122–147.

Bandura, A. (1989). Human agency in social cognitive theory. *American Psychologist, 44,* 1175–1184.

Bardo, P. (1979). The pain of teacher burnout: A case history. *Phi Delta Kappan, 61,* 252–254.

Barnett, P. A., & Gotlib, I. H. (1988). Psychosocial functioning and depression. *Psychological Bulletin, 93,* 97–126.

Barron, F. (1957). Originality in relation to personality and intellectual ability. *Journal of Personality, 25,* 58–74.

Barron, F. (1963a). *Creativity and psychological health.* Princeton, NJ: Van Nostrand.

Barron, F. (1963b). The needs for order and for disorder as motives in creative activity. In C. W. Taylor & F. Barron (Eds.), *Scientific creativity: Its recognition and development* (pp. 153–160). New York: John Wiley and Sons.

Barron, F. (1972). *Artists in the making.* New York: Seminar Press.

Bartley III, W. W. (1983). *Wittgenstein, ein Leben* [Wittgenstein, a life]. München: Matthes & Seitz.

Beakley, G. C., & Leach, H. W. (1967). *Engineering: An introduction to a creative profession.* New York: Macmillan.

Beck, A. T. (1976). *Cognitive therapy and emotional disorders.* New York: International Universities Press.

Beck, A. T., & Young, J. E. (1985). Depression. In D. H. Barlow (Ed.), *Clinical handbook of psychological disorders* (pp. 206–244). New York: Guilford Press.

Becker, E. (1973). *The denial of death.* New York: Free Press.

Beehr, T. A. (1981). Work-role stress and attitudes to co-workers. *Group & Organizational Studies, 6,* 201–210.

Beehr, T. A., & Bhagat, R. S. (1985). *Human stress and cognition in organizations: An integrated perspective.* New York: John Wiley and Sons.

Beittell, K. R. (1964). Creativity in the visual arts in higher education. In C. W. Taylor (Ed.), *Widening horizons in creativity* (pp. 379–395). New York: John Wiley and Sons.

Bejat, M. (1975). Cognitive style and personality traits in scientific creativity. *Revue Roumaine des Sciences Sociales, 2,* 129–142.

Belcastro, P. A., Gold, R. S., & Hays, L. C. (1983). Maslach Burnout Inventory: Factor structures for samples of teachers. *Psychological Reports, 53,* 364–366.

Berlyne, D. E. (1965). *Structure and direction in thinking.* New York: John Wiley and Sons.

Betz, N. E., & Hackett, G. (1986). Applications of self-efficacy theory to understanding career choice behavior. *Journal of Social and Clinical Psychology, 4,* 279–289.

Bibeau, G., Dussault, G., Larouche, L. M., Lippel, K., Saucier, J. F., Vézina, M., & Vidal, J. M. (1989). *Certains aspects culturels, diagnostiques et juridiques de burnout* [Some cultural diagnostic and juridical aspects of burnout]. Montréal: Confédération des Syndicats Nationaux.

Birch, N. E., Marchant, M. P., & Smith, N. M. (1986). Perceived role conflict, role ambiguity, and reference librarian burnout in public libraries. *Library & Information Science Journal, 8,* 53–65.

Blalock, S. J., McEvoy-DeVellis, B., & DeVellis, R. F. (1989). Social comparison among individuals with rheumatoid arthritis. *Journal of Applied Social Psychology, 19,*(8), 665–680.

Bloch, A. M. (1976). The battered teacher. *Today's Education, 66,* 58–62.

Blostein, S., Eldridge, W., Kilty, K., & Richardson, V. (1985). A multidimensional analysis of the concept of burnout. *Employee Assistance Quarterly, 1,* 55–66.

Bode, K. (1988). *Burnout bei Hamburger Krankenschwestern* [Burnout in Hamburg nurses]. Unpublished master's thesis, University of Hamburg.

* Boudreau, R. A., & Golembiewski, R. T. (1989). Burnout and stress in American, Canadian, and Japanese work settings: Nomothetic and ideographic perspectives. *Kaihatsu Ronshu, 44,* 53–77.

* Boudreau, R. A., & Golembiewski, R. T. (1990). Modes of response to advanced burnout: Note on a Japanese urban health-care population. *Kaihatsu Ronshu, 45,* 37–53.

Bowlby, J. (1979). *The making and breaking of affectional bonds*. London: Tavistock.

Boy, A. V., & Pine, G. J. (1980). Avoiding counselor burnout through role renewal. *Personnel & Guidance Journal, 59,* 161–163.

Bramhall, M., & Ezell, S. (1981). How burned-out are you? *Public Welfare, 39*(1), 23–27.

Bräutigam, W. (1969). *Reaktionen, Neurosen, Psychopathien* [Reactions, neuroses, psychopathies]. München: Deutscher Taschenbuchverlag.

Brickman, P., Rabinowitz, B. C., Karuza, J., Cohn, E., & Kidder, L. (1982). Models of helping and coping. *American Psychologist, 37,* 368–384.

Brickman, P., & Bulman, R. J. (1977). Pleasure and pain in social comparison. In J. M. Suls & R. L. Miller (Eds.), *Social comparison processes* (pp. 149–186). New York: Hemisphere.

Brill, P. L. (1984). The need for an operational definition of burnout. *Family & Community Health, 6,* 12–24.

Brookings, J. B., Bolton, B., Brown, C. E., & McEvoy, A. (1985). Self-reported job burnout among female human service professionals. *Journal of Occupational Behavior, 6,* 143–150.

Brown, G. W., & Harris, T. (1978). *Social origins of depression*. London: Free Press.

Bryman, A. (1986). *Leadership in organizations*. London: Routledge.

Bühl, W. L. (1984). *Krisentheorien* [Theories of crisis]. Darmstadt: Wissenschaftliche Buchgemeinschaft.

* Burden, R. L. (1988). Stress and the school psychologist. *School Psychology International, 9,* 55–59.

Burisch, M. (1984a). Approaches to personalty inventory construction. *American Psychologist, 39,* 214–227.

Burisch, M. (1984b). *The Maslach Burnout Inventory and the Tedium Measure: Reliability and validity in a German sample*. Unpublished manuscript, University of Hamburg.

Burisch, M. (1989). *Das Burnout-Syndrom. Theorie der inneren Erschöpfung* [The burnout syndrome. A theory of inner exhaustion]. Heidelberg: Springer.

Burke, R. J. (1984). Beliefs and fears underlying type A behavior. *Psychological Reports, 54,* 655–662.

Burke, R. J. (1987). Issues and implications for health care delivery systems: A Canadian perspective. In J. C. Quick, R. S. Bhagat, J. E. Dalton, and J. D. Quick (Eds.), *Work stress* (pp. 34–50). New York: Praeger.

Burke, R. J. (1989a). Toward a phase model of burnout. *Group & Organization Studies, 14,* 23–32.

Burke, R. J. (1989b). Sources of managerial and professional stress in large organizations. In C. L. Cooper & R. Payne (Eds.), *Causes, coping and consequences of stress at work* (pp. 77–114). Chichester: John Wiley and Sons.

Burke, R. J., & Deszca, G. (1986). Correlates of psychological burnout phases among police officers. *Human Relations, 39,* 487–501.

Burns, T., & Stalker, G. M. (1961). *The management of innovation.* London: Tavistock.

Buunk, B. P. (1990). Affiliation and helping interactions within organizations: A critical analysis of the role of social support with regard to occupational stress. In W. Stroebe & M. Hewstone (Eds.), *European review of social psychology* (Vol. 1, pp. 293–322). Chichester: John Wiley and Sons.

Buunk, B. P., Collins, R., VanYperen, N. W., Taylor, S. E., & Dakoff, G. (1990). Upward and downward comparisons: Either direction has its up and downs. *Journal of Personality & Social Psychology, 59,* 1238–1249.

Buunk, B. P., Schaufeli, W. B., & Ybema, J. F. (1990). *Occupational burnout: A social comparison perspective.* Paper presented at the ENOP Conference on Professional Burnout, Krakow, Poland.

Buunk, B. P., Schaufeli, W. B, & Ybema, J. F. (1992). *Social comparison and professional burnout.* Manuscript submitted for publication.

Buunk, B. P., & VanYperen, N. W. (1991). Referential comparisons, relational comparisons and exchange orientation: Their relation to marital satisfaction. *Personality & Social Psychology Bulletin, 17,* 710–718.

Buunk, B. P., VanYperen, N. W., Taylor, S. E., & Collins, R. B. (1991). The drive upward in social comparison theory revisited. *European Journal of Social Psychology, 21,* 529–546.

Byrne, B. M. (1991). The Mastach Burnout Inventory: Validating factorial and invariance across intermediate, secondary and university educators. *Multivariate Behavioral Research, 26,* 583–605.

Caccese, C. F., & Mayerberg, C. K. (1984). Gender differences in perceived burnout of college coaches. *Journal of Sport Psychology, 6,* 279–288.

Cahoon, A. R., & Rowney, J. I. (1984). Managerial burnout: A comparison by sex and level of responsibility. *Journal of Health & Human Resources Administration, 7,* 249–263.

Camus, A. (1955). *The myth of sisyphus.* New York: Random House.

Capel, S. A., Sisley, B. L., & Desertrain, G. S. (1987). The relationship of role conflict and role ambiguity to burnout in high school basketball coaches. *Journal of Sport Psychology, 9,* 106–117.

Caplan, G. (1964). *Principles of preventive psychiatry.* New York: Basic Books.

Caron, C., Corcoran, K. J., & Simcoe, F. (1983). Intrapersonal correlates of burnout: The role of locus of control and self-esteem. *The Clinical Supervisor, 14,* 53–62.

Carrière, J. (1987). *Le prix d'un Goncourt* [The price of a Goncourt prize]. Paris: Editions R. Laffont.

Cherniss, C. (1980a). *Professional burnout in the human service organizations.* New York: Praeger.

Cherniss, C. (1980b). *Staff burnout: Job stress in the human services.* Beverly Hills: Sage.

Cherniss, C. (1989a). Burnout in new professionals: A long-term follow-up study. *Journal of Health & Human Resources Administration, 12,* 11–24.

Cherniss, C. (1989b). Career stability in public service professionals: A longitudinal investigation based on biographical interviews. *American Journal of Community Psychology, 17,* 399–422.

Cherniss, C. (1990). *Organizational negotiation skill and the prevention of burnout: Lessons of a long-term follow-up study.* Paper presented at the 98th Annual Convention of the American Psychological Association, Boston.

Cherniss, C. (1992). Long-term consequences of burnout: An exploratory study. *Journal of Organizational Behavior, 13,* 1–11.

Cherniss, C., & Krantz, D. L. (1983). The ideological community as an antidote to burnout in the human services. In B. Farber (Ed.), *Stress and burnout in the human service professions* (pp. 198–212), New York: Pergamon Press.

Childers, J. H. (1985). Organizational health: How to measure a school's level of health and take remedial action. *Journal of Educational Public Relations, 8,* 4–7.

Clark, D. (1990). *A study of stress in Canadian Forces personnel employed in traditional and non-traditional environments.* Unpublished master's thesis, Acadia University, Wolfville, Canada.

Clark, M. S., Ouellette, R., Powell, M. C., & Milberg, S. (1987). Recipient's mood, relationship type, and helping. *Journal of Personality & Social Psychology, 53,* 93–103.

Cobb, S. (1976). Social support as a moderator of life stress. *Psychosomatic Medicine, 38,* 300–314.

Cobb, S., & Rose, R. H. (1973). Hypertension, peptic ulcers, and diabetes. *Journal of the Australian Medical Association, 224,* 489–492.

* Cohen, J. (1976). German and American workers: A comparative view of worker distress. *International Journal of Mental Health, 5,* 130–147.

Cohen, S., & Hoberman, H. M. (1983). Positive events and social supports as buffers of life change stress. *Journal of Applied Social Psychology, 13,* 99–125.

Constable, J. F., & Russell, D. W. (1986). The effect of social support and the work environment upon burnout among nurses. *Journal of Human Stress, 12,* 20–26.

Cooper, A. M. (1986). Narcissism. In A. P. Morrison (Ed.), *Essential papers on narcissism.* (pp. 122–143). New York: New York University Press.

* Cooper, C. L., & Arbose, J. (1984). Executive stress goes global. *International Management, 39,* 42–48.

* Cooper, C. L., & Hensman, R. (1985). A comparative investigation of executive stress: A ten-nation study. *Stress Medicine, 1,* 295–301.

Cooper, C. L., & Payne, R. (Eds.). (1988). *Causes, coping and consequences of stress at work.* Chichester: John Wiley and Sons.

Corbin, A. (1990). Backstage. In M. Perrot (Ed.), *A history of private life: Vol. 4. From the fires of revolution to the great war* (pp. 451–668). Cambridge, MA: Harvard University Press.

Corcoran, K. J. (1986). Measuring burnout: A reliability and convergent validity study. *Journal of Social Behavior & Personality, 1,* 107–112.

Cottrell, N. B., & Epley, S. W. (1977). Affiliation, social comparison, and socially mediated stress reduction. In J. M. Suls & R. L. Miller (Eds.). *Social comparison processes* (pp. 43–68). New York: Hemisphere.

Cox, T. (1978). *Stress.* London: Macmillan.

Cox, T. (1985). The nature and measurement of stress. *Ergonomics, 28,* 1155–1163.

Cox, T. (1988a) Cognitive science, occupational stress and organizational health. *Work & Stress, 2,* 193–198.

Cox, T. (1988b). Organizational health. *Work & Stress, 2,* 1–2.

Cox, T. (1988c). Psychobiological factors in stress and health. In S. Fisher, & J. Reason (Eds.), *Handbook of life stress, cognition and health* (pp. 603–628). Chichester: John Wiley and Sons.

Cox, T. (1990). The recognition and measurement of stress: (conceptual and methodological issues. In J. R. Wilson & N. Corlett (Eds.), *Evaluation of human work* (pp. 628–647). London: Taylor & Francis.

Cox, T, Cox, S., Farnsworth, B., & Boot, N. with Walton, C., & Ferguson, E. (1989). *Teachers and schools: A study of organizational health and stress.* Unpublished manuscript, University of Nottingham.

Cox, T., Davis, A., & Cook, D. (1991). *The effects of repetitive computer-based work on performance and well-being.* Unpublished manuscript, University of Nottingham.

Cox, T., & Ferguson, E. (1991). Individual differences, stress and coping. In C. L. Cooper and R. Payne (Eds.), *Personality and stress* (pp. 7–30). Chichester: John Wiley and Sons.

Cox, T., & Howarth, I. (1990). Organizational health, culture and helping. *Work & Stress, 4,* 107–110.

Cox, T., & Kuk, G. (1991). *Healthiness of schools as organizations: Teacher stress and health.* Paper presented at the International Congress on Stress, Anxiety and Emotional Disorders, University of Minho, Braga, Portugal.

Cox. T., Kuk, G., & Schur, H. (1991). *The meaningfulness of work to professional burnout.* Unpublished manuscript, University of Nottingham.

Cox, T., Leather, P., & Cox, S. (1990). Stress, health and organizations. *Occupational Health Review*, February/March, 13–18.

Cox, T., & Mackay, C. J. (1985). The measurement of self-reported stress and arousal. *British Journal of Psychology, 76,* 183–186.

Cox, T., Thirlaway, M., & Cox, S. (1984). Occupational well being: Sex differences at work. *Ergonomics, 27,* 499–510.

Cox, T., Thirlaway, M., Gotts, G., & Cox, S. (1983). The nature and measurement of general well being. *Journal of Psychosomatic Research, 27,* 353–359.

Coyne, J. C. (1976a). Depression and response of others. *Journal of Abnormal Psychology, 85,* 186–193.

Coyne, J. C. (1976b). Toward an interactional description of depression. *Psychiatry, 39,* 28–40.

Craig, T. K. J., & Brown, G. W. (1984). Goal frustration and life events in the aetiology of painful gastrointestinal disorder. *Journal of Psychosomatic Research, 28,* 411–421.

Cronin-Stubbs, D., & Rooks, C. (1985). The stress, social support, and burnout of critical care nurses. *Heart & Lung, 14,* 31–39.

Crowne, D. P., & Marlowe, D. (1964). *The approval motive.* Chichester: John Wiley and Sons.

Crutchfield, R. S. (1962). Conformity and creative thinking. In H. E. Gruber, G. Terrel, & M. Wertheimer (Eds.), *Contemporary approaches to creative thinking* (pp. 120–140). New York: Atherton Press.

Daily, A. L. (1985). The burnout test. *American Journal of Nursing, 85,* 270–272.

Davis-Sacks, M. L., Jayaratne, S., & Chess, W. A. (1985). A comparison of the effects of social support on the incidence of burnout. *Social Work, 30,* 240–244.

Deci, E. L. (1975). *Intrinsic motivation.* New York: Plenum Press.

* DeFrank, R. S., Ivancevich, J. M., & Schweiger, D. M. (1988). Job stress and mental well-being: Similarities and differences among American, Japanese, and Indian managers. *Behavioral Medicine, 14,* 160–170.

De Wolff, Ch. J. (1986). Stress and strain in the work environment: Does it lead to illness? In W. D. Gentry, H. Benson, & Ch. J. de Wolff (Eds.), *Behavioral medicine: Work, stress and health.* Dordrecht: Martinus Nijhoff.

Dignam, J. T., Barrera, M., & West, S. G. (1986). Occupational stress, social support, and burnout among correctional officers. *American Journal of Community Psychology, 14,* 177–193.

Dignam, J. T., & West, S. G. (1988). Social support in the workplace: Tests of six theoretical models. *American Journal of Community Psychology, 16,* 701–724.

Dohrenwend, B. S., Dohrenwend, B. P., Dodson, M., & Shrout, P. E. (1984).

Symptoms, hassles, social supports and life events: The problem of confounded measures. *Journal of Abnormal Psychology, 93,* 222–230.

Dolan, A. (1987). The relationship between burnout and job satisfaction in nurses. *Journal of Advanced Nursing, 12,* 3–12.

Dollard, J., Doob, L. W., Miller, N. E., Mowrer, O. H., Sears, R. R., Ford, C. S., Hovland, C. I., & Sollenberger, R. T. (1939). *Frustration and aggression.* New Haven: Yale University Press.

Drory, A., & Shamir, B. (1988). Effects of organizational and life variables on job satisfaction and burnout. *Group & Organization Studies, 13,* 441–455.

Dunkel-Schetter, C. (1984). Social support and cancer: Findings based on patient interviews and their implications. *Journal of Social Issues, 40,* 77–98.

Eckenrode, J., & Gore, S. (Eds.). (1990). *Stress between work and families.* New York: Plenum Press.

Edelwich, J., & Brodsky, A. (1980). *Burn-out: Stages of disillusionment in the helping professions.* New York: Human Sciences Press.

Einsiedel, A. A., & Tully, H. A. (1982). Methodological considerations in studying burnout. In J. W. Jones (Ed.), *The burnout syndrome: Current research, theory, interventions* (pp. 89–106), Park Ridge, IL: London House.

Eisenstat, R. A., & Felner, R. D. (1984). Toward a differentiated view on burnout: Personal and organizational mediators of job satisfaction and stress. *American Journal of Community Psychology, 12,* 411–430.

Eliasz, A. (1980). Temperament and trans-situational stability of behavior in the physical and social environment. *Polish Psychological Bulletin, 11,* 143–153.

Emener, W. G., Luck, R. S., & Gohs, F. X. (1982). A theoretical investigation of the construct burnout. *Journal of Rehabilitation Administration, 6,* 188–196.

England, G. W. (1988). *The patterning of work meanings which are conterminous with work outcome levels for individuals in Japan, Germany, and the U.S.A..* Unpublished manuscript.

England, G. W., & Whitely, W. T. (1992). Cross-national meanings of working. In A. Brief & W. Nord (Eds.), *The meaning of work.* (pp. 65–106) Lexington, MA: Lexington Books.

Enzmann, D., & Kleiber, D. (1989). *Helfer-Leiden: Stress und Burnout in psychosozialen Berufen* [Helpers' ordeals: Stress and burnout in human services professions]. Heidelberg: Asanger-Verlag.

Etzion, D. (1987). *Burnout: The hidden agenda of human distress* (Working Paper No. 930/87). Tel-Aviv: Tel-Aviv University.

* Etzion, D., Kafry, D., & Pines, A. (1982). Tedium among managers: A cross-cultural American–Israeli comparison. *Journal of Psychology & Judaism, 7,* 30–41.

* Etzion, D., & Pines, A. (1986). Sex and culture in burnout and coping among

human service professionals. *Journal of Cross-Cultural Psychology, 17,* 191–209.

* Etzion, D., Pines, A., & Kafry, D. (1983). Coping strategies and the experience of tedium: A cross-cultural comparison between Israelis and Americans. *Journal of Psychology & Judaism, 8,* 41–51.

Evans, B. K., & Fischer, D. (1989). *The dimensionality and discriminate validity of the Maslach Burnout Inventory.* Paper presented at the Annual convention of the Canadian Psychological Association, Halifax, Nova Scotia, Canada.

* Evans, G. W., Palsane, M. N., & Carrere, S. (1987). Type A behavior and occupational stress: A cross-cultural study of blue-collar workers. *Journal of Personality & Social Psychology, 52,* 1002–1007.

Ezrahi, Ch. (1987). *What is burnout? The relationships among subjective burnout, objective burnout and personality traits.* Unpublished master's thesis, School of Education, Hebrew University, Jerusalem.

Facaoaru, C., & Macarie, E. (1976). Factorial analysis of a set of creativity tests. *Revue Roumaine des Sciences Sociales, 2,* 145–158.

Farber, B. (1982). *Stress and burnout: Implications for teacher motivation.* Paper presented at the Annual Meeting of the American Educational Research Association, New York.

Farber, B. (1983a). Dysfunctional aspects of the psychotherapeutic role. In B. Farber (Ed.), *Stress and burnout in the human service professions* (pp. 97–118), New York: Pergamon Press.

Farber, B. (1983b). Introduction: A critical perspective on burnout. In B. Farber (Ed.), *Stress and burnout in the human service professions* (pp. 1–22). New York: Pergamon Press.

Farber, B. A. (1984). Stress and burnout in suburban teachers. *Journal of Educational Research, 77,* 325–331.

Feldman, S. P. (1988). How organizational culture can affect innovation. *Organizational Dynamics, 17,* 57–68.

Fender, L. K. (1989). Athlete burnout: Potential for research and intervention strategies. *Sport Psychologist, 3,* 63–71.

Festinger, L. (1954). A theory of social comparison processes. *Human Relations, 7,* 117–140.

Fiedler, F. E., & Garcia, J. E. (1987). *New approaches to effective leadership: Cognitive resources and organizational performance.* Chichester: John Wiley and Sons.

Fimian, M. J. (1984). Organizational variables related to stress and burnout in community based programs. *Education & Training of the Mentally Retarded, 19,* 201–209.

Fimian, M. J., & Blanton, L. P. (1987). Stress, burnout, and role problems among teacher trainees and first year teachers. *Journal of Occupational Behavior, 8,* 157–165.

Fimian, M. J., Fastenau, P. A., Tashner, J. H., & Cross, A. H. (1989). The measure of classroom stress and burnout among gifted and talented students. *Psychology in the Schools, 26,* 139–153.

Firth, J. (1985). Personal meaning of occupational stress: Cases from the clinic. *Journal of Occupational Psychology, 58,* 139–148.

Firth, H., & Britton, P. (1989). "Burnout," absence and turnover amongst British nursing staff. *Journal of Occupational Psychology, 62,* 55–60.

Firth, H., McIntee, J., & McKeown, P. (1985). Maslach Burnout Inventory: Factor structure and norms for British nursing staff. *Psychological Reports, 57,* 147–150.

Firth, H., McIntee, J., McKeown, P., & Britton, P. G. (1986a). Burnout and professional depression: Related concepts? *Journal of Advanced Nursing, 11,* 633–641.

Firth, H., McIntee, J., McKeown, P., & Britton, P. G. (1986b). Interpersonal support among nurses at work. *Journal of Advanced Nursing, 11,* 273–282.

Fischer, H. J. (1983). A psychoanalytic view of burnout. In B. A. Farber (Ed.), *Stress and burnout in the human service professions* (pp. 40–46). New York: Pergamon Press.

Fiske, D. W. (1973). How shall we conceptualize the personality we seek to investigate. In J. R. Royce (Ed.), *Multivariate analysis and psychological theory* (pp. 75–102). London: Academic Press.

Fletcher, B. (1988). The epidemiology of occupational stress. In C. L. Cooper & R. Payne, *Causes, coping and consequences of stress at work,* (pp. 3–52), Chichester: John Wiley and Sons.

Ford, D. L., Murphy, C. J., & Edwards, K. L. (1983). Exploratory development and validation of a Perceptual Job Burnout Inventory. *Psychological Reports, 52,* 995–1006.

Forney, D. S., Wallace-Schutzman, F., & Wiggers, T. T. (1982). Burnout among career development professionals: Preliminary findings and implications. *Personnel & Guidance Journal, 60,* 435–439.

Frank, J. D. (1961). *Persuasion and healing.* Baltimore: Johns Hopkins Press.

Frank, R. (1989). *Burnout bei Schwesternschülerinnen* [Burnout in nursing students]. Unpublished master's thesis, University of Hamburg.

Frankl, V. (1963). *Man's search for meaning.* Boston: Beacon Press.

Freedman, R. (1978). *Hermann Hesse. Pilgrim of crisis.* New York: Pantheon.

Freedy, J. R. (1990). *Stress inoculation for prevention of burnout: A conservation of resources approach.* Unpublished doctoral dissertation, Kent State University, Kent, Ohio.

Freedy, J. R., Shaw, D. L., Jarrell, M. P., & Masters, C. R. (1992). Toward an understanding of the psychological impact of natural disasters: An application of the conservation of resources stress model. *Journal of Traumatic Stress, 5,* 441–454.

French, J. R. P., & Caplan, R. D. (1970). Psychosocial factors in coronary heart disease. *Industrial Medicine, 39,* 383–397.

French, J. R. P., & Caplan, R. D. (1972). Organizational stress and individual strain. In A. J. Marrow (Ed.), *The Failure of Success.* New York: AMA-COM.

Freudenberger, H. J. (1974). Staff burnout. *Journal of Social Issues, 30,* 159–165.

Freudenberger, H. J. (1975). The staff burnout syndrome in alternative institutions. *Psychotherapy: Theory, Research, & Practice, 12,* 72–83.

Freudenberger, H. J. (1981). *Burnout: Contemporary issues and trends.* Paper presented at the National Conference on Stress and Burnout, New York.

Freudenberger, H. J. (1983a). Burnout: Contemporary issues, trends, and concerns. In B. A. Farber (Ed.), *Stress and burnout in the human service professions* (pp. 23–28). New York: Pergamon Press.

Freudenberger, H. J., & North, G. (1986). *Att inte räcka till* [Women's burnout]. Köping: Bonniers.

Freudenberger, H. J., & Richelson, G. (1980). *Burnout: The high cost of high achievement.* Garden City, NY: Doubleday.

Friedman, I. (1991). *High and low burnout schools: Sources of stress at the classroom and school-level.* Paper presented at the Conference on Teacher Stress and Burnout, Teachers College, Columbia University, New York.

Gaines, J., & Jermier, J. M. (1983). Emotional exhaustion in a high stress organization. *Academy of Management Journal, 26,* 567–586.

Ganster, D. C. (1987). Type A behavior and occupational stress. *Journal of Organizational Behavior Management, 8,* 61–84.

Ganster, D. C. (1989). Worker control and well-being: A review of research in the workplace. In S. L. Sauter, J. J. Hurrell, & C. Cooper, *Job control and worker health* (pp. 3–24). Chichester: John Wiley and Sons.

Garden, A. M. (1987a). Depersonalization: A valid dimension of burnout? *Human Relations, 40,* 545–560.

Garden, A. M. (1987b). Demographic characteristics and personality types of resident assistants as predictor variables of job satisfaction, burnout, or supervisory rating. *Dissertation Abstracts International, 49,* 1701–A.

Gaudinski, M. A. (1982). Coping with expending nursing practice, knowledge and technology. In E. A. McConnell (Ed.), *Burnout in the nursing profession: Coping strategies, causes, and costs* (pp. 81–85). St. Louis, MO: Mosby.

Gentry, W. D., Foster, S. B., & Froehling, S. (1972). Psychological response to situational stress in intensive and non-intensive nursing. *Heart & Lung, 1,* 793–796.

Gerard, H. B. (1963). Emotional uncertainty and social comparison. *Journal of Abnormal & Social Psychology, 66,* 586–573.

Gerstein, L. H., Topp, C. G., & Correll, G. (1987). The role of the environment and person when predicting burnout among correctional personnel. *Criminal Justice & Behavior, 14,* 352–369.

Gibbons, F. X., & Gerrard, M. (1991). Downward comparison and coping with threat. In J. Suls & T. A. Wills (Eds.), *Social comparison: Contemporary theory and research* (pp. 317–346). Hillsdale, NJ: Erlbaum.

Gibson, S. & Dembo, M. H. (1984). Teacher efficacy: A construct validation. *Journal of Educational Psychology, 76,* 569–582.

Gil-Monte, P. R., & Schaufeli, W. B. (1992). Burnout en enermería: Un estudio compatativo Espana–Holanda (Burnout among nurses: A comparative Spanish–Dutch study]. *Revista de Psicología del Trabajo y de las Organizaciones, 7,* 121–130.

Ginsburg, S. G. (1974). The problem of the burned out executive. *Personnel Journal, 53,* 598–600.

Girault, N. (1989). *Burnout: Emergence et strategies d'adaption* [Burnout: Emergence and strategies of adaptation]. Unpublished doctoral dissertation, Université René Descartes, Paris.

Gmelch, W. H., Lovrich, N. P., & Wilke, P. K. (1983). A national study of stress among university faculty members. *Phi Delta Kappan, 65,* 367.

Gold, Y. (1984). The factorial validity of the Maslach Burnout Inventory in a sample of California elementary and junior high school classroom teachers. *Educational & Psychological Measurement, 44,* 1009–1016.

Gold, Y., Bachelor, P., & Michael, W. B. (1989). The dimensionality of a modified form of the MBI for university students in a teacher training program. *Educational & Psychological Measurement, 49,* 549–561.

Goldberger, L., & Breznitz, S. (Eds.). (1982). *Handbook of stress.* New York: Free Press.

Golembiewski, R. T., Boudreau, R. A., Goto, K., & Murai, T. (1992). *Transnational perspectives on burnout: Is the phase model generic or culturally-bounded in a Japanese replication?* Paper presented at the International Conference on Advances in Management, Orlando, FL.

* Golembiewski, R. T., Boudreau, R. A., Kim, B-S., Munzenrider, R. F., & Park, S.-J. (1990). *Two aspects of burnout in cross-cultural settings: Are the phase model and its underlying MBI sub-domains generic?.* Paper presented at the ENOP Conference on Professional Burnout, Kraków, Poland.

Golembiewski, R. T., Hilles, R., & Daly, R. (1987). Some effects of multiple OD interventions on burnout and worksite features. *Journal of Applied Behavioral Science, 23,* 295–314.

Golembiewski, R. T., & Munzenrider, R. F. (1988). *Phases of burnout: Developments in concepts and applications.* New York: Praeger.

Golembiewski, R. T., Munzenrider, R., & Carter, D. (1983). Phases of progressive burnout and their work site covariates. *Journal of Applied Behavioral Science, 19,* 461–481.

Golembiewski, R. T., Munzenrider, R. F., & Stevenson, J. G. (1986). *Stress in organizations: Toward a phase model of burnout*. New York: Praeger.

Gomes, M. E., & Maslach, C. (1991). *Commitment and burnout among political activists: An in-depth study*. Paper presented at the 14th Annual Meeting of the International Society of Political Psychology, Helsinki, Finland.

Goudy, E., & Spielberger, Ch. P. (1975). Influence of anxiety on the effectiveness of learning. *Journal of Educational Psychology, 5*, 78–93.

Gough, H. G., & Woodworth, D. W. (1960). Stylistic variations among professional research scientists. *Journal of Psychology, 49*, 87–98.

Gray-Toft, P., & Anderson, J. G. (1981). Stress among hospital nursing staff: Its causes and effects. *Social Science & Medicine, 15*, 639–647.

* Green, D. E., & Walkey, F. H. (1988). A confirmation of the three-factor structure of the Maslach Burnout Inventory. *Educational & Psychological Measurement, 48*, 579–585.

Green, D. E., Walkey, F. H., & Taylor, A. J. W. (1991), The three-factor structure of the Maslach Burnout Inventory. *Journal of Social Behavior & Personality, 6*, 453–472.

Greene, G. (1960). *A burnt out case*. London: Heinemann.

Greenglass, E. R. & Burke, R. J. (1988). Work and family precursors of burnout in teachers: Sex differences. *Sex Roles, 18*, 215–229.

Grey, J. A. (1972). Learning theory, the conceptual nervous system and personality. In V. D. Nebylitsyn & J. A. Grey (Eds.), *Biological bases of individual behavior* (pp. 134–152). New York: Academic Press.

Guilford, J. P. (1967). *The nature of human intelligence*. New York: McGraw-Hill.

Haack, M., & Jones, J. W. (1983). Diagnosing burnout using projective drawings. *Journal of Psychosocial Nursing & Mental Health Services, 21*, 9–16.

Hackman, J. R. (1986). The psychology of self-management in organizations. In M. S. Pallak & R. Perloff (Eds.), *Psychology and work: Productivity, change, and employment* (pp. 123–161). Washington, DC: American Psychological Association.

Hall, D. T. (1971). A theoretical model of career subidentity development in organizational settings. *Organizational Behavior & Human Performance, 66*, 50–76.

Hall, D. T. (1976). *Careers in organizations*. Pacific Palisades, CA: Goodyear.

Hallsten, L. (1985). *Burnout: En studie kring anpassnings- och utvecklingsprocesser i en byråkrati* [Burnout: A study on adaptive- and developmental processes in a bureaucracy], Stockholm: Länsarbetsnämnden.

Hallsten, L. (1986). *Burnout in a Swedish bureaucracy: Some data and a model*. Paper presented at the NIVA course on Occupational Stress and Staff Burnout among Human Service Personnel, Helsinki, Finland.

Hallsten, L. (1988). *En organisation möter utbrändhet—ett försök med rådslag* [An organization encounters burnout], No. 11. IPSO, Stockholm.

Hammer, J. S., Jones, J. W., Lyons, J. S., Sixmith, J. S., & Afficiando, R. N. (1985). Measurement of occupational stress in hospital settings. *General Hospital Psychiatry, 7,* 156–162.

Handy, C. B. (1985). *Understanding organizations.* London: Penguin Books.

Handy, J. A. (1988). Theoretical and methodological problems within occupational stress and burnout research. *Human Relations, 41,* 351–369.

Hansson, R. O., Jones, W. H., & Carpenter, B. N. (1984). Relationship competence and social support. In P. Shaver (Ed.), *Review of personality and social psychology* (Vol. 5, pp. 265–284). Beverly Hills, CA: Sage.

* Harari, H., Jones, C. A., & Sęk, H. (1988). Stress syndromes and stress predictors in American and Polish college students. *Journal of Cross-Cultural Psychology, 19,* 243–255.

Harrison, D. H. (1983). A social competence model of burnout. In B. Farber (Ed.), *Stress and burnout in the human service professions* (pp. 29–39). New York: Pergamon Press.

Hatfield, E., Cacioppo, J. T., & Rapson, R. (1992). In M. S. Clark (Ed.). *Review of personality and social psychology.* (vol. 14, pp. 151–177) Newbury Park, CA: Sage.

Heifetz, L. J., & Bersani, H. A. (1983). Disrupting the cybernetics of personal growth: Toward a unified theory of burnout in the human services. In B. A. Farber (Ed.), *Stress and burnout in the human service professions* (pp. 46–64). New York: Pergamon Press.

Heimler, E. (1975). *Survival in society.* London: Weidenfeld & Nicolson.

Heimler, E. (1985). *The healing echo.* London: Souvenir Press.

* Himle, D. P., Jayaratne, S., & Thyness, P. (1986). Predictors of job satisfaction, burnout and turnover among social workers in Norway and the USA: A cross-cultural study. *International Social Work, 29,* 323–334.

Hobfoll, S. E. (1988). *The ecology of stress.* New York: Hemisphere.

Hobfoll, S. E. (1989). Conservation of resources. A new attempt at conceptualizing stress. *American Psychologist, 44,* 513–524.

Hobfoll, S. E., & Lieberman, Y. (1987). Personality and social resources in immediate and continued stress resistance among women. *Journal of Personality & Social Psychology, 52,* 18–26.

Hobfoll, S. E., & Lerman, M. (1988). Personal relationships, personal attitudes, and stress resistance: Mothers' reactions to their child's illness. *American Journal of Community Psychology, 16,* 565–589.

Hobfoll, S. E., & Lerman, M. (1989). Predicting receipt of social support: A longitudinal study of parents' reactions to their child's illness. *Health Psychology, 8,* 61–77.

Hobfoll, S. E., Lilly, R. S., & Jackson, A. P. (1992). Conservation of social resources and the self. In H. O. F. Veiel & U. Baumann (Eds.), *The mean-*

ing and measurement of social support: Taking stock of 20 years of research (pp. 125–141) Washington, DC: Hemisphere.

Holley, J. W., & Guilford, J. P. (1964). A note on the G index of agreement. *Educational & Psychological Measurement, 24*, 749–753.

Hoy, W. K., & Feldman, J. A. (1987). Organizational health: The concept and its measure. *Journal of Research & Development in Education, 20*, 30–37.

Huberty, T. J., & Huebner, E. S. (1988). A national survey of burnout among school psychologists. *Psychology in the Schools, 25*, 54–61.

Ipser, K. (1987). *Mit Goethe in Italien. Eine historische Reise* [Goethe in Italy. A historical journey]. Herrsching: Pawlak.

Iwanicki, E. F., & Schwab, R. L. (1981). A cross validation study of the Maslach Burnout Inventory. *Educational & Psychological Measurement, 41*, 1167–1174.

Jackson, S. E. (1983). Participation in decision-making as a strategy for reducing job-related strain. *Journal of Applied Psychology, 68*, 3–19.

Jackson, S. E. (1984). Organizational practices for preventing burnout. In A. Sethi & R. Schulder (Eds.), *Handbook of organizational stress coping strategies*. Cambridge, MA: Ballinger.

Jackson, S. E., & Maslach, C. (1982). After-effects of job-related stress: Families as victims. *Journal of Occupational Behavior, 3*, 63–77.

Jackson, S. E., Schwab, R. L., & Schuler, R. S. (1986). Toward an understanding of the burnout phenomenon. *Journal of Applied Psychology, 71*, 630–640.

Jones, J. W. (1980a). *Preliminary manual: The Staff Burnout Scale for health professionals*. Park Ridge, IL: London House Press.

Jones, J. W. (1980b). *A measure of staff burnout among health professionals*, Paper presented at the Annual Convention of the American Psychological Association, Montreal, Quebec, Canada.

Jöreskog, K. G., & Sörbom, D. (1985). *LISREL VI: Analysis of linear structural relationships by the method of maximum likelihood*. Mooresville, NJ: Scientific Software.

Justice, B., Gold, R. S., & Klein, J. P. (1981). Life events and burnout. *Journal of Psychology, 108*, 219–226.

Kadushin, A. (1974). *Child welfare services*. New York: Macmillan.

Kahill, S. (1986). Relationship of burnout among professional psychologists to professional expectations and social support. *Psychological Reports, 59*, 1043–1051.

Kahill, S. (1988). Symptoms of professional burnout: A review of the empirical evidence. *Canadian Psychology, 29*, 284–297.

Kahn, R. L., Wolfe, D. M., Quinn, R. P., Snoek, J. D., & Rosenthal, R. A. (1964). *Organizational stress. Studies in role conflict and ambiguity*. New York: John Wiley and Sons.

Kanner, A. D., Coyne, J. C., Schaefer, C., & Lazarus, R. S. (1981). Comparisons of two modes of stress measurement: Daily hassles and uplifts versus major life events. *Journal of Behavioral Medicine, 4,* 1–39.

Kanner, A., Kafry, D., & Pines, A. (1978). Conspicuous in its absence: The lack of positive conditions as a source of stress. *Journal of Human Stress, 4,* 33–39.

Kanungo, R. N. (1979). The concept of alienation and involvement revisited. *Psychological Bulletin, 86,* 119–138.

* Karasek, R. A. (1979). Job demands, job decision latitude, and mental strain: Implications for job redesign. *Administrative Science Quarterly, 24,* 285–308.

Karasek, R. (1989). Control in the workplace and its health-related aspects. In S. L. Sauter, J. J. Hurrell, & C. L. Cooper (Eds.), *Job control and worker health* (pp. 129–159). Chichester: John Wiley and Sons.

Kasl, S. V. (1980). The impact of retirement. In C. L. Cooper & R. Payne (Eds.), *Current concerns in occupational stress* (pp. 137–186). Chichester: John Wiley and Sons.

Katz, D. (1964). The motivational basis of organizational behavior. *Behavioral Science, 9,* 131–146.

Katz, S., & Mazur, M. A. (1979). *Understanding the rape victim: A synthesis of research findings.* Chichester: John Wiley and Sons.

Kegan, R. (1982). *The evolving self: Problem and process in human development.* Cambridge, MA.: Harvard University Press.

Keinan, G., & Melamed, S. (1987). Personality characteristics and proneness to burnout: A study among internists. *Stress Medicine, 3,* 307–315.

* Keinan, G., & Perlberg A. (1987). Stress in academe. *Journal of Cross-Cultural Psychology, 18,* 193–207.

Kets de Vries, M. F. R., & Miller, D. (1984). *The neurotic organization.* London: Jossey-Bass.

Kilpatrick, A. O. (1986). *Burnout: An empirical assessment.* Unpublished doctoral dissertation, University of Georgia, Athens.

Kim, B.-S. (1990). *Alternative models of burnout phases.* Unpublished doctoral dissertation, University of Georgia, Athens.

Kimpson, R. S., & Sonnabend, L. C. (1978). Public schools: The interrelationships between organizational health and innovativeness and between organizational health and staff characteristics. *Urban Education, 10,* 27–48.

Kirton, M. J. (1976). Adaptors and innovators: A description and measure. *Journal of Applied Psychology, 61,* 622–629.

Kirton, M. J. (1977). *Manual of the Kirton Adaption-Innovation Inventory.* London: National Foundation for Educational Research.

Kirton, M. J. (1980). Adaptors and innovators: The way people approach problems. *Planned Innovation, 3,* 51–54.

Kirton, M. J. (1991). Adaptors and innovators: Why new initiatives get blocked. In J. Henry (Ed.), *Creative management* (pp. 209–220). London: Sage.

Kleiber, D., & Enzmann, D. (1986). Helfer-Leiden: Überlegungen zum Burnout in helfenden Berufen &Helpers' ordeals: On burnout in the helping professionsé. In G. Feuser & W. Jantzen (Eds.), *Jahrbuch für Psychopathologie und Pychotherapie [Yearbook of psychopathology and psychotherapy]*, (Vol. 6, pp. 49–78). Köln: Pahl-Rugenstein.

Kleiber, D., & Enzmann, D. (1990). *Burnout: 15 years of research: An international bibliography.* Göttingen: Hogrefe.

Klinger, E. (1975). Consequences of commitment to and disengagement from incentives. *Psychological Review, 82,* 1–25.

Klinger, E. (1977). *Meaning and void. Inner experience and the incentives in peoples lives.* Minneapolis: University of Minnesota Press.

Kobasa, S. C. (1979). Stressful life events, personality, and health: An inquiry into hardiness. *Journal of Personality & Social Psychology, 37,* 1–11.

Kobasa, S. C., Maddi, S. R., & Zola, M. A. (1983). Type A and hardiness. *Journal of Behavioral Medicine, 6,* 41–51.

Koeske, G. F., & Koeske, R. D. (1989). Construct validity of the Maslach Burnout Inventory: A critical review and reconceptualization. *Journal of Applied Behavioral Science, 25,* 131–144.

Kohut, H. (1971). *The analysis of the self: A systematic approach to the psychoanalytic treatment of narcissistic personality disorders.* New York: International Universities Press.

Kramer, M. (1974). *Reality shock.* St. Louis, MO: C.V. Mosby.

Kyriacou, C. (1987). Teacher stress and burnout: An international review. *Educational Research, 22,* 146–152.

Lahoz, M., & Mason, L. (1989). Maslach Burnout Inventory: Factor structures and norms for USA pharmacists. *Psychological Reports, 64,* 1059–1063.

Lammers, C. J. (1983). *Organisaties vergelijkenderwijs* [Organizations compared]. Utrecht: Spectrum.

Landsbergis, P. A. (1988). Occupational stress among health care workers: A test of the job-demands model. *Journal of Organizational Behavior, 9,* 217–239.

Lang, D. (1985). Preconditions of three types of alienation in young managers and professionals. *Journal of Occupational Behavior, 6,* 171–182.

Langer, E. J. (1979). The illusion of incompetence. In L. C. Perlmuter & R. A. Monty (Eds.), *Choice and perceived control.* Hillsdale, NJ: Erlbaum.

Lasch, C. (1979). *The culture of narcissism,* New York: Warner.

Latack, J. C. (1984). Career transitions within organizations: An exploratory study of work, nonwork, and coping strategies. *Organizational Behavior & Human Performance, 34,* 296–322.

Lauderdale, M. (1982). *Burnout.* Austin, TX: Learning Concepts.

Laughlin, H. P. (1967). *The neuroses.* Washington, DC: Butterworths.

Lazaro, L., Shinn, M., & Robinson, P. E. (1985). Burn out, performance and job withdrawal behavior. *Journal of Health & Human Resources Administration, 7,* 213–234.

Lazarus, R. S., DeLongis, A., Folkman, S., & Gruen, R. (1985). Stress and adaptational outcomes: The problem of confounded measures. *American Psychologist, 40,* 770–779.

Lazarus, R. S., & Folkman, S. (1984). *Stress, appraisal, and coping.* New York: Springer.

Lee, R. T. & Asforth, B. E. (in press). A longitudinal study of burnout among supervisors and managers: Comparisons between Leiter and Maslach (1988) and Golembiewski et al. (1986) models. *Organizational Behavior and Human Decision Processes.*

Lee, R. T., & Ashforth, B. E. (1990). On the meaning of Maslach's three dimensions of burnout. *Journal of Applied Psychology, 75,* 743–747.

Lefcourt, H. M., (1976). *Locus of control: Current trends in theory and research.* New York: Halstead.

Leiter, M. P. (1988a). Burnout as a function of communication patterns: A study of a multidisciplinairy mental health team. *Group & Organizational Studies, 13,* 111–128.

Leiter, M. P. (1988b). Commitment as a function of stress reactions among nurses: A model of psychological evaluations of worksettings. *Canadian Journal of Community Mental Health, 7,* 115–132.

Leiter, M. P. (1989). Conceptual implications of two models of burnout: A response to Golembiewski. *Group & Organizational Studies, 14,* 15–22.

Leiter, M. P. (1990a). *Burnout as a crisis in professional role development.* Paper presented at the ENOP Conference on Professional Burnout, Kraków, Poland.

Leiter, M. P. (1990b). The impact of family resources, control coping and skill utilization on the development of burnout: A longitudinal study. *Human Relations, 43,* 1067–1083.

Leiter, M. (1991a). Coping patterns as predictors of burnout. *Journal of Organizational Behavior, 12,* 123–144.

Leiter, M. (1991b). The dream denied: Professional burnout and the constraints of service organizations. *Canadian Psychology, 32,* 547–558.

Leiter, M. P. (1992). Burnout as a crisis in self-efficacy: Conceptual and practical implications. *Work & Stress 6,* 107–115.

Leiter, M. P., & Maslach, C. (1988). The impact of interpersonal environment on burnout and organizational commitment. *Journal of Organizational Behavior, 9,* 297–308.

Leiter, M. P., & Meechan, K. A. (1986). Role structure and burnout in the field of human services. *Journal of Applied Psychology, 22,* 47–52.

Lerner, M. J. (1980). *The belief in a just world: A fundamental delusion.* New York: Plenum Press.

Levinson, H. (1981). When executives burn out. *Harvard Business Review, 59,* 73–81.

Lewin, K. (1936). The psychology of success and failure. *Occupations, 14,* 926–930.

Lief, H. I., & Fox, R. C. (1963). Training for "detached concern" in medical students. In H. I. Lief, V. F. Lief, & N. R. Lief (Eds.), *The psychological basis of medical practice* (pp. 12–35). New York: Harper & Row.

Lindquist, C. A., & Whitehead, J. T. (1986). Burnout, job stress and job satisfaction among southern correctional officers. *Journal of Offender Counseling, Services & Rehabilitation, 10,* 5–26.

Linville, P. W. (1985). Self-complexity and affective extremity: Don't put all your eggs in one cognitive basket. *Social Cognition, 3,* 94–120.

Loevinger, J. (1976). *Ego development: Conceptions and theories.* San Francisco: Jossey-Bass.

Mackay, C., Cox, T., Burrows, G., & Lazzerini, T. (1978). An inventory for the measurement of self-reported stress and arousal. *British Journal of Social & Clinical Psychology, 17,* 283–284.

MacKinnon, D. W. (1964). Creativity and images of the self. In R.W. White (Ed.), *The study of lives* (pp. 251–278). New York: Atherton Press.

Maddi, S. (1967). The existential neurosis. *Journal of Abnormal Psychology, 72,* 311–325.

Maddi, S. (1970). The search for meaning. In W. Arnold & M. Page (Eds.), *The Nebraska symposium on motivation.* (pp. 137–186). Lincoln: University of Nebraska Press.

Major, B., Testa, M., & Bylsma, W. H. (1991). Responses to upward and downward social comparisons. In J. Suls & T. A. Wills (Eds.), *Social comparison* (pp. 237–260). Hillsdale, NJ: Erlbaum.

Mandler, G. (1991). *In the absence of the sacred.* San Francisco: Sierra Club.

Mann, T. (1922). *Buddenbrooks: Verfall einer Familie* [Buddenbrooks: Decline of a family]. Berlin: S. Fischer (1st ed. 1900).

* Manso-Pinto, J. F. (1989). Occupational stress factors as perceived by Chilean school teachers. *Journal of Social Psychology, 129,* 127–129.

Marcelissen, F. H. G. (1987). *Gangmakers van het stressproces.* [Psychological pacemakers in the stress process]. Leiden: NIPG/TNO.

Marek, T., Fąfrowicz, M., & Noworol, C. (1993). Stimulation seeking and potentially creative activity. In T. Marek (Ed.), *Psychological mechanisms of human creativity: The temptation for reassessment* (pp. 166–177). Delft: Eburon.

Maslach, C. (1973). *"Detached concern" in health and social service professions.* Paper presented at the annual meeting of the American Psychological Association, Montreal, Quebec, Canada.

Maslach, C. (1976). Burned-out. *Human Behavior, 5,* 16–22.

Maslach, C. (1978a). The client role in staff burnout. *Journal of Social Issues, 34,* 111–124.

Maslach, Ch. (1978b). Job burnout: How people cope, *Public Welfare, 36,* 56–58.

Maslach, C. (1979). Negative emotional biasing of unexplained arousal. *Journal of Personality & Social Psychology, 37,* 953–969.

Maslach, C. (1982a). *Burnout: The cost of caring.* Englewood Cliffs, NJ: Prentice-Hall.

Maslach, C. (1982b). Burnout: A social psychological analysis. In J. W. Jones (Ed.), *The burnout syndrome: Current research, theory, investigations* (pp. 30–53). Park Ridge, IL: London House Press.

Maslach, C. (1982c). Understanding burnout: Definitional issues in analyzing a complex phenomenon. In W. S. Paine (Ed.), *Job stress and burnout* (pp. 29–40). Beverly Hills, CA: Sage.

Maslach, C., & Jackson, S. E. (1981a). The measurement of experienced burnout. *Journal of Occupational Behavior, 2,* 99–113.

Maslach, C., & Jackson, S. E. (1981b). *The Maslach Burnout Inventory. Research edition.* Palo Alto, CA: Consulting Psychologists Press.

Maslach, C., & Jackson, S. E. (1982). Burnout in health professions: A social psychological analysis. In G. Sanders & J. Suls (Eds.), *Social psychology of health and illness* (pp. 227–251). Hillsdale, NJ: Erlbaum.

Maslach, C., & Jackson, S. E. (1984a). Burnout in organizational settings. In S. Oskamp (Ed.), *Applied social psychology annual 5* (pp. 133–154). Beverly Hills, CA: Sage.

Maslach, C., & Jackson, S. E. (1984b). Patterns of burnout among a national sample of public contact workers. *Journal of Health & Human Resources Administration, 7,* 189–212.

Maslach, C., & Jackson, S. E. (1985). The role of sex and family variables in burnout. *Sex Roles, 12,* 837–851.

Maslach, C., & Jackson, S. E. (1986). *The Maslach Burnout Inventory. Manual* (2nd ed.). Palo Alto, CA: Consulting Psychologists Press.

Maslach, C., & Pines, A. (1977). The burnout syndrome in the day care setting. *Child Care Quarterly, 6,* 100–113.

Maslow, A. H. (1959). Creativity in self-actualizing people. In H. H. Anderson (Ed.), *Creativity and its cultivation* (pp. 83–95). New York: Harper.

Matthews, K. A. (1982). Psychological perspectives on the type A behavior pattern. *Psychological Bulletin, 91,* 293–323.

* McCormick, I. A., & Cooper, C. L. (1988). Executive stress: Extending the international comparison. *Human Relations, 41,* 65–72.

McCranie, E. W., & Brandsma, J. M. (1988). Personality antecedents of burnout among middle-aged physicians. *Behavioral Medicine, 14,* 30–36.

McCranie, E. W., Lambert, V., & Lambert, C. (1987). Work stress, hardiness and burnout among hospital staff nurses. *Nursing Research, 36,* 374–378.

McGinnes, J. (1987). Free radicals and the developmental pathology of schizo-phrenic burnout. *Integrative Psychiatry, 5,* 288–301.

McGoldrick, A. F., & Cooper, C. L. (1985). Stress at the decline of one's career. In T. A. Beehr & R. S. Bhagat (Eds.), *Human stress and cognition in organizations* (pp. 177–201). Chichester: John Wiley and Sons.

McGrath, A., Reid, N., & Boore, J. Occupational stress in nursing. *International Journal of Nursing Studies, 26,* 343–358.

McMullen, M. B., & Krantz, M. (1988). Burnout in day care workers: The effects of learned helplessness and self-esteem. *Child & Youth Care Quarterly, 17,* 275–280.

McQueen, D. V., & Siegrist, J. (1982). Social factors in the etiology of chronic disease: An overview. *Social Science & Medicine, 16,* 353–367.

Meehl, P. E. (1978). Theoretical risks and tabular asterisks: Sir Karl, Sir Ronald, and the slow progress of soft psychology. *Journal of Consulting & Clinical Psychology, 46,* 806–834.

Meier, S. T. (1983). Towards a theory of burnout. *Human Relations, 36,* 899–910.

Meier, S. T. (1984). The construct validity of burnout. *Journal of Occupational Psychology, 57,* 211–219.

Meier, S. T., & Schmeck, R. R. (1985). The burned out college student: A descriptive profile. *Journal of College Students Personnel, 26,* 63–69.

Melamed, S., Kushnir, T., & Shirom, A. (1992). *Burnout and risk factors for cardiovascular diseases. Behavioral Medicine, 18,* 53–60.

Messic, S. (1969). Measures of cognitive styles and personality and their potential for educational practice. In K. Ingenkamp (Ed.), *Developments in educational testing* (pp. 329–341). London: University of London Press.

Messic, S. (1973). Multivariate models of cognition and personality. In J. R. Royce (Ed.), *Multivariate analysis and psychological theory* (pp. 265–303). London: Academic Press.

Miles, M. B. (1965). Planned change and organizational health: Figure and ground. In R. O. Carlson, A. Gallaher, M. B. Miles, R. J. Pellegrin, & E. M. Rogers (Eds.), *Change processes in the public schools* (pp. 11–36). Eugene, OR: Center for the Advanced Study of Educated Administration.

Miller, D. T., & McFarland, C. (1991). When social comparison goes awry: The case of pluralistic ignorance. In J. Suls & T. A. Wills (Eds.), *Social comparison,* (pp. 287–316). Hillsdale, NJ: Erlbaum.

Miller, K. I., Stiff, J. B., & Ellis, B.H. (1988). Communication and empathy as precursors to burnout among human service workers. *Communication Monographs, 55,* 250–265.

Miller, N., Gross, S., & Holtz, R. (1991). Social projection and attitudinal certainty. In J. Suls & T. A. Wills (Eds.), *Social comparison* (pp. 177–210). Hillsdale, NJ: Erlbaum.

Mills, J. & Mintz, P. M. (1972). Effect of unexplained arousal on affiliation. *Journal of Personality & Social Psychology, 24,* 11–13.

Mintzberg, H. (1979). *The structuring of organizations.* Englewood Cliffs, NJ.: Prentice-Hall.

Mohler, S. R. (1983). The human element in air traffic control. *Aviation, Space, & Environmental Medicine, 54,* 511–516.

Molleman, E., Pruyn, J., & Van Knippenberg, A. (1986). Social comparison processes among cancer patients. *British Journal of Social Psychology, 25,* 1–13.

Monat, A., & Lazarus, R. S. (1985). *Stress and coping.* New York: Columbia University Press.

Mor, V., & Laliberte, L. (1984). Burnout among hospice staff. *Health & Social Work, 9,* 274–283.

Moritz, M. (1991). *Arbeitszufriedenheit und Burnout. Eine empirische Untersuchung* [Job satisfaction and burnout. An empirical study]. Unpublished master's thesis, University of Hamburg.

Morrison, A. P. (Ed.). (1986). *Essential papers on narcissism.* New York: New York University Press.

Moscovici, S. (1985). Social influence and conformity. In G. Lindzey & E. Aronson (Eds.), *Handbook of social psychology* (pp. 347–412). New York: Random House.

Mulkay, M. S. (1972). *The social process of innovation.* London: Macmillan.

Munton, A. G., & Forster, N. (1990). Job relocation: Stress and the role of the family. *Work & Stress, 4,* 75–81.

Murphy, L. R. (1988). Workplace interventions for stress reduction and prevention. In C. L. Cooper & R. Payne (Eds.), *Causes, coping and consequences of stress at work* (pp. 301–342). Chichester: John Wiley and Sons.

Murstein, B. I., Cerreto, M., & MacDonald, M. G. (1977). A theory and investigation of the effect of exchange-orientation on marriage and friendships. *Journal of Marriage & the Family, 39,* 543–548.

Nadler, A. (1991). Help-seeking behavior: Psychological costs and instrumental benefits. In M. S. Clark (Ed.), *Review of personality and social psychology* (Vol. 12, pp. 290–311). Newbury Park, CA: Sage.

Nagi, S., & Davis, L. G. (1985) Burnout: A comparative analysis of personality and environmental variables. *Psychological Reports, 57,* 1319–1328.

Nosal, C. S. (1979). *Mechanizmy funkcjonowania intelektu: zdolnosci, style poznawcze, przetwarzanie informacji* [Mechanisms of performing intellect: Abilities, cognitive styles, processing information]. Wroclaw: Wroclaw University of Technology Press.

Nowakowska, M. (1970). Nieformalne ujecie wspolczesnej teorii testow [Informal formulation of theory]. *Studia Sociologiczne, 3,* 245–273.

Oatley, K., & Bolton, W. (1985). A social-cognitive theory of depression in reaction to life events. *Psychological Review, 92,* 372–388.

O'Driscoll, M. P., & Schubert, T. (1988). Organizational climate and burnout in a New Zealand social service agency. *Work and Stress, 2,* 199–204.

* Orth-Gomer, K. (1979). Ischemic heart disease and psychological stress in Stockholm and New York. *Journal of Psychosomatic Research, 23,* 165–173.

Orthner, D., & Pittman, J. (1986). Family contributions to work commitment. *Journal of Marriage & Family Living, 48,* 573–581.

Oswin, M. (1978). *Children living in long stay hospitals.* London: Heinemann.

Ozer, E. M., & Bandura, A. (1990). Mechanisms governing empowerment effects: A self-efficacy analysis. *Journal of Personality & Social Psychology, 58,* 472–486.

Paine, W. S. (1982a). The burnout syndrome in context. In J.W. Jones (Ed.), *The burnout syndrome: Current research, theory, interventions* (pp. 1–29), Park Ridge, IL: London House Press.

Paine, W. S. (1982b). Overview: Burnout stress syndromes and the 1980's. In W. S. Paine (Ed.), *Job stress and burnout* (pp. 11–25), Beverly Hills, CA: Sage.

Payne R. L., & Jones, J. G. (1987). Measurement and methodological issues in social support. In S. V. Kasl & C. L. Cooper (Eds.), *Stress and health: Issues in research methodology* (pp. 167–206). Chichester: John Wiley and Sons.

Pelsma, D. M., Roland, B., Tollefson, N., & Wigington, H. (1989). Parent burnout: Validation of the MBI with a sample of mothers. *Measurement and Evaluation in Counseling and Development, 22,* 81–87.

Penn, M., Romano, J. L., & Foat, D. (1988). The relationship between job satisfaction and burnout. *Administration in Mental Health, 15,* 157–165.

Perlman, B., & Hartman, A. E. (1982). Burnout: Summary and future research. *Human Relations, 35,* 283–305.

Pick, D., & Leiter, M. (1991). Nurses' perceptions of the nature and causes of burnout: A comparison of self reports and standardized measures. *Canadian Journal of Nursing Research, 23,* 33–48.

Pierce, C. M., & Molloy, G. N. (1989). The construct validity of the MBI: Some data from down under. *Psychological Reports, 65,* 1340–1342.

Pines, A. (1981). Burnout: A current problem in pediatrics. *Current Issues in Pediatrics,* May issue.

Pines, A. M. (1982a). Changing organizations: Is a work environment without burnout an impossible goal? In W. S. Paine (Ed.), *Job stress and burnout: Research, theory, intervention perspectives* (pp. 189–211). London: Sage.

Pines, A. (1982b). Helpers' motivation and the burnout syndrome. In T. A. Wills (Ed.), *Basic processes in helping relationships* (pp. 453–475). New York: Academic Press.

Pines, A. M. (1988). *Keeping the spark alive: Preventing burnout in love and marriage.* New York: St. Martin's Press.

Pines, A. (in press, b). Burnout in political activism: An existential perspective. *Journal of Health & Human Resources Administration.*

Pines, A. (1992). A burnout workshop: Design and rationale. In R. T. Golembiewski (Ed.), *Handbook of organizational consultation* (pp. 615–619). New York: Marcel Dekker.

Pines, A (in press, d). The Palistinean Intifada and Isreali's burnout. *Journal of Cross Cultural Psychology.*

Pines, A., & Aronson, E. (1988). *Career burnout: Causes and cures.* New York: Free Press.

* Pines, A., Aronson, E., & Kafry, D. (1981). *Burnout: From tedium to personal growth.* New York: Free Press.

* Pines, A., & Caspi, A. (1992). Causes of burnout in organizational consultation. In R. T. Golembiewski (Ed.), *Handbook of organizational consultation* (pp. 615–619). New York: Marcel Dekker.

Pines, A., & Kafry, D. (1978). Occupational tedium in social service professionals. *Social Work, 23,* 499–507.

* Pines, A., Kafry, D., & Etzion D. (1980). Job stress from a cross-cultural perspective. In K. Reid (Ed.) *Burnout in the helping professions* (pp. 1–13). Kalamazoo, MI: Western Michigan University Press.

Pines, A., & Kanner, A. (1982). Nurses' burnout: Lack of positive conditions and presence of negative conditions as two independent sources of stress. *Journal of Psychosocial Nursing, 8,* 30–35.

Pines, A., & Maslach, C. (1978). Characteristics of staff burnout in mental health settings. *Hospital & Community Psychiatry, 29,* 233–237.

Pines, A., & Maslach, C. (1980). Combatting staff burnout in a day care center: A case study. *Child Care Quarterly, 9,* 5–16.

Pittner, M. S., & Houstone, B. K. (1980). Response to stress, cognitive coping strategies and the type A behavior pattern. *Journal of Personality & Social Psychology, 39,* 147–157.

Pnina, S., & Leiner, K. (1975). Creativity and Anxiety. *Psychology in the Schools, 3,* 69–76.

Powers, S., & Gose, K. F. (1986). Reliability and construct validity of the Maslach Burnout Inventory in a sample of university students. *Educational & Psychological Measurement, 46,* 251–255.

Procaccini, J., & Kiefaber, M. W. (1983). *Parent burnout.* Garden City, NY: Doubleday.

Pyszcynsky, T., & Greenberg, J. (1987). Self-regulatory perseveration and depressive self-focusing style. *Psychological Bulletin, 102,* 122–138.

Rabbie, J. M. (1963). Differential preference for companionship under threat. *Journal of Abnormal & Social Psychology, 67,* 643–648.

Rafferty, J. P., Lemkau, J. P., Purdy, R. R., & Rudisill, J. R. (1986). Validity of the Maslach Burnout Inventory for family practice physicians. *Journal of Clinical Psychology, 42,* 488–492.

Reiter, L., & Strotzka, H. (1977). Der Begriff der Krise [The concept of crisis]. *Psychiatria Clinica, 10,* 7–26.

Reykowski, J. (1978). Osobowosc jako centralny system regulacji i integracji czynnosci [Personality as a central system of activities' regulation and integration]. In T. Tomaszewski (Ed.), *Psychologia* (p. 750). Warszawa: PWN.

Roberts, C. A. (1986). Burnout: Psycho-babble or a valuable concept? *British Journal of Hospital Medicine, 10,* 63–66.

Rofé, Y. (1984). Stress and affiliation. A utility theory. *Psychological Review, 91,* 235–250.

Rogers, E. H. (1960). *The ecology of health.* New York: Macmillan.

* Romo, M., Siltanen, P., Theorell, T., & Rahe, R. H. (1974). Work behavior, time urgency and life dissatisfactions in subjects with myocardial infarction: A cross-cultural study. *Journal of Psychosomatic Research, 18,* 1–8.

Rosch, P. J., & Pelletier, K. R. (1989). Designing worksite stress-management programs. In L. R. Murphy & T. F. Schoenborn (Eds.), *Stress management in work settings* (pp. 65–86). New York: Praeger.

Rosenberg, M. (1979). *Conceiving the self.* New York: Basic Books.

Ross, L., Rodin, J., & Zimbardo, P. G. (1969). Toward an attribution therapy: The reduction of fear through induced cognitive-emotional misattribution. *Journal of Personality & Social Psychology, 12,* 279–288.

Rossiter, A. Jr. (1979). Burnout syndrome peril for professionals, executives. *The Knickerbockers News,* Tuesday, May 1.

Rotter, J. B. (1966). Generalized expectancies for internal control versus external control of reinforcement. *Psychological Monographs, 80,* 1–28.

* Rowney, J. I. A., & Cahoon, A. R. (1990). *Cross-cultural characteristics of burnout.* Proceedings of the International Management Conference (pp. 211–222). Xi'an, China.

Royce, J. R. (1973). The conceptual framework for a multi-factor theory of individuality. In J. R. Royce (Ed.), *Multivariate analysis and psychological theory* (pp. 305–407). London: Academic Press.

Sarnoff, I., & Zimbardo, P. G. (1961). Anxiety, fear, and social facilitation. *Journal of Abnormal & Social Psychology, 62,* 597–605.

Savicki, V. Cooley, E. J. (1983). Theoretical and research considerations of burnout. *Children and Youth Services Review, 5,* 227–238.

Schachter, S. (1959). *The psychology of affiliation.* Palo Alto, CA: Stanford University Press.

Schachter, S., & Singer, J. (1962). Cognitive, social, and physiological determinants of the emotional state. *Psychological Review, 69,* 379–399.

Schaufeli, W. B. (1990). *Opgebrand* [Burnout]. Rotterdam: Ad. Donker.

Schaufeli, W. B., & Buunk, B. P. (1991). *Burnout and social evaluation processes in organizations.* Paper presented at the 5th European Congress on the Psychology of Work and Organization, Rouen, France.

Schaufeli, W. B., & Janczur, B. (in press). Burnout among nurses: A Polish–Dutch comparison. *Journal of Cross-Cultural Psychology.*

Schaufeli, W. B., & Peeters, M. C. M. (1990). *The measurement of burnout.* Paper presented at the ENOP Conference on Professional Burnout, Kraków, Poland.

Schaufeli, W. B., & Van Dierendonck, D. (1992). *The construct validity of two burnout measures.* Submitted for publication.

Schein, E. H. (1972). *Professional education: Some new directions.* New York: McGraw-Hill.

Schmidbauer, W. (1977). *Die hilflosen Helfer* [Helpless helpers]. Reinbek: Rowohlt.

Schönpflug, W. (1985). Goal directed behavior as a source of stress. In M. Frese & J. Sabini (Eds.), *The concept of action in psychology* (pp. 172–188). Hillsdale, NJ: Erlbaum.

Schuler, R. S. (1980). Definition and conceptualization of stress in organizations. *Organizational Behavior & Human Performance, 23,* 184–215.

Schwab, R. L., & Iwanicki, E. F. (1982). Perceived role conflict, role ambiguity, and teacher burnout. *Educational Administration Quarterly, 18,* 60–74.

Schwartz, M. S., & Will, G. T. (1953). Low morale and mutual withdrawal on a hospital ward. *Psychiatry, 16,* 337–353.

Schwarzer, R., & Leppin, A. (1989). *Socialer Rückhalt und Gesundheit—Eine Meta-Analyse* [Social support and health—A meta analysis]. Göttingen: Hogrefe.

Seeman, M. (1967). On the personal consequences of alienation in work. *American Sociological Review, 32,* 273–285.

Segal, Z. V. (1988). Appraisal of the self-schema construct in cognitive models of depression. *Psychological Bulletin, 103,* 147–162.

Selye, H. (1967). *Stress in health and disease.* Boston: Butterworth.

Selye, H. (1951–1956). *Annual report of stress.* New York: McGraw-Hill.

Selye, H. (1956). *The stress of life.* New York: McGraw-Hill.

Selye, H. (1982). History and present status of the stress concept. In L. Goldberger & S. Breznitz (Eds.), *Handbook of stress* (pp. 7–20). New York: Free Press.

Sexton, P. (1982). *The new nightingales.* New York: Enquiry.

Shamir, B. (1986). *Some arguments against the use of burnout as a broadly applied developmental variable.* Paper presented at the 21st Congress of the International Association of Applied Psychology, Jerusalem.

Shepard, J. M. (1972). Alienation as a process: Work as a case in point. *The Sociological Quarterly, 13,* 161–173.

Shirom, A. (1986). *Does stress lead to affective strain, or vice versa? A struc-*

tural regression test. Paper presented at the 21st Congress of the International Association of Applied Psychology, Jerusalem.

Shirom, A. (1989). Burnout in work organizations. In C. L. Cooper & I. Robertson (Eds.), *International review of industrial and organizational psychology* (pp. 25–48), Chichester: John Wiley and Sons.

Shouksmith, C. (1970). *Intelligence, creativity and cognitive style.* London: Batsford.

* Shouksmith, G., & Burrough, S. (1988). Job stress factors for New Zealand and Canadian air traffic controllers. *Applied Psychology: An International Review, 37,* 263–270.

Shubin, S. (1978). Burnout: The professional hazard you face in nursing. *Nursing, 8,* 22–27.

Silver, R. L., & Wortman, C. B. (1980). Coping with undesirable life events. In I. Garber & M. Seligman (Eds.), *Human helplessness: Theory and applications* (pp. 279–375). New York: Academic Press.

Singer, J. A., Neale, M. S., & Schwartz, G. E. (1989). The nuts and bolts of assessing occupational stress. In L. R. Murphy & T. F. Schoenborn (Eds.), *Stress management in work settings* (pp. 3–30). New York: Praeger.

Sirigatti, S., Stefanile, C., & Menoni, E. (1988). Per un adattamento italiano del Maslach Burnout Inventory [About the Italian adaptation of the MBI]. *Belletino de Piscologia Applicata, 187–188,* 33–39.

Skelton, J. A., & Pennebaker, J. W. (1982). The psychology of physical symptoms and sensations. In G. S. Sanders & J. Suls (Eds.), *Social psychology of health and illness* (pp. 99–128). Hillsdale, NJ: Erlbaum.

Smith, R. E. (1986). Toward a cognitive-affective model of athletic burnout. *Journal of Sport Psychology, 8,* 36–50.

Smith, R. H., & Insko, C. A. (1987). Social comparison choice during ability evaluation. *Personality & Social Psychology Bulletin, 13,* 111–120.

Solomon, Z., Mikulincer, M., & Hobfoll, S. E. (1986). Effects of social support and battle intensity on loneliness and breakdown in combat. *Journal of Personality & Social Psychology, 51,* 1269–1276.

Starlie, F. J. (1982). Burnout: The elaboration of a concept. In E. A. McConnell (Ed.) *Burnout in the nursing profession: coping strategies, causes, and costs* (pp. 81–85). St. Louis, MO: Mosby.

Stephens, M. A. P., Norris, V. K., Kinney, J. M., Ritchie, S. W., & Grotz, R. C. (1988). Stressful situations in caregiving: Relationships between caregiving, coping and well-being. *Psychology & Aging, 3,* 208–209.

Steward, M. S., & Maszaros, P. S. (1981). What's your burnout-score?. *Journal of Home Economics, 73,* 37–39.

Stokols, D. (1975). Toward a psychological theory of alienation. *Psychological Review, 82,* 26–44.

Stolorow, R. D. (1986). Toward a functional definition of narcissism. In A. P.

Morrison (Ed.), *Essential papers on narcissism* (pp. 197–210). New York: New York University Press.

Storms, M. D., & Nisbett, R. E. (1970). Insomnia and the attribution process. *Journal of Personality & Social Psychology, 16,* 319–328.

Stout, J. K., & Williams, J. M. (1983) Comparison of two measures of burnout. *Psychological Reports, 53,* 283–289.

Strelau, J. (1983). *Temperament, personality, activity.* New York: Academic Press.

Strzalecki, A. (1969). *Wybrane zagadnienia psychologii tworczosci* [Selected problems of psychology of creativity]. Warszawa: PWN.

Strzalecki, A. (1989). *Tworczosc a style rozwiazywania problemow. Ujecie prakseologiczne* [Creativity and styles of solving practical problems. Praxeological issues]. Wroclaw: Ossolineum.

Sullins, E. S. (1991). Emotional contagion revisited: Effects of social comparison and expressive style on mood convergence. *Personality & Social Psychology Bulletin, 17,* 166–174.

Swann, W. B., Jr., & Brown, J. D. (1990). From self to health: Self-verification and identity disruption. In B. R. Sarason, I. G. Sarason, & G. R. Pierce (Eds.), *Social support: An interactional view* (pp. 150–172). Chichester: John Wiley and Sons.

Tarter, C. J., Hoy, W. K., & Kottkamp, R. B. (1990). School health and organizational commitment. *Journal of Research & Development in Education, 23,* 236–242.

Taylor, C. (1989). *Sources of the self: The making of the modern identity.* Cambridge, MA: Cambridge University Press.

Taylor, S., Buunk, B. P., & Aspinwall, L. (1990). Social comparison, stress and coping. *Personality & Social Psychology Bulletin, 16,* 74–89.

Taylor, S. E., & Lobel, M. (1989). Social comparison activity under threat: Downward evaluation and upward contacts. *Psychological Review, 96,* 569–575.

Teichmann, Y. (1987). Affiliation and self disclosure in a specific ego threat situation. *Personality & Individual Differences, 8,* 807–812.

Thoits, P. A. (1983b). Multiple identities and psychological well-being: A reformulation and test of the social isolation hypothesis. *American Sociological Review, 48,* 174–187.

Thoits, P. A. (1983). Dimensions of life events that influence psychological distress. In H. B. Kaplan (Ed.), *Psychosocial stress: Trends in theory and research* (pp. 33–103). New York: Academic Press.

* Tokar, E., & Feitler, F. C. (1986). A comparative study of teacher stress in American and British middle schools. *Journal of Early Adolescence, 6,* 77–82.

Turnipseed, D. L. (1987). Burnout among hospice nurses: An empirical assesssment. *Hospice Journal, 3,* 105–119.

Tversky, A., & Kahneman, D. (1981). The framing of decisions and the psychology of choice. *Science, 24,* 453–458.

Tymstra, T. (1989). The imperative character of medical technology and the meaning of "anticipated decision regret." *International Journal of Technology Assessment in Health Care, 5,* 207–213.

Ulich, D. (1987). *Krise und Entwicklung. Zur Psychologie der seelischen Gesundheit* [Crisis and development. On the psychology of mental health]. München: Psychologie Verlags Union.

Ursprung, A. W. (1986). Incidence and correlates of burnout in residential service settings. *Rehabilitation Counseling Bulletin, 29,* 225–239.

Van Dijkhuizen, N. (1980). *From stressors to strains.* Lisse: Swets & Zeitlinger.

VanYperen, N. W., Buunk, B. P., & Schaufeli, W. B. (1992). Imbalance, communal orientation and the burnout syndrome among nurses. *Journal of Applied Social Psychology, 22,* 173–189.

Veninga, R. L., & Spradley, J. P. (1981). *The work/stress connection: How to cope with job burnout.* Boston: Little, Brown.

Veyne, P. (Ed.). (1987). *A history of private life (Vol. 1).* Cambridge, MA: Harvard University Press.

Victor, B., & Cullen, J. B. (1988). The organizational bases of ethical work climates. *Administrative Science Quarterly, 33,* 101–125.

Viney, L. L. (1976). The concept of crisis: A tool for clinical psychologists. *Bulletin of the British Psychological Society, 29,* 387–395.

Wade, D. C., Cooley, E., & Savicki, V. (1986). A longitudinal study of burnout. *Children & Youth Services Review, 8,* 161–173.

Wallace, M., Levens, L., & Singer, G. (1988). Blue collar stress. In C. L. Cooper, & R. Payne (Eds.), *Causes, coping and consequences of stress at work* (pp. 53–76). Chichester: John Wiley and Sons.

Walster, E., Berscheid, E., & Walster, G. W. (1978). *Equity: Theory and research.* Boston: Allyn and Bacon.

Warr, P. B. (1987). *Work, unemployment and mental health.* Oxford: Clarendon Press.

Warr, P. B. (1990). The measurement of well being and other aspects of mental health. *Journal of Occupational Psychology, 63,* 193–210.

Watson, D., & Pennebaker, J. W. (1989). Health complaints, stress, and distress: Exploring the central role of negative affectivity. *Psychological Review, 96,* 234–254.

Weber, M. (1970). *Essays in sociology.* London: Routledge.

Weinberg, S., Edwards, G. M., & Garove, W. E. (1983). Burnout among employees of state residential facilities serving developmentally disabled persons. *Children & Youth Services Review, 5,* 239–253.

Wheeler, L. (1974). Social comparison and selective affiliation. In T. Hudson

(Ed.), *Foundations of interpersonal attraction* (pp. 309–330). New York: Academic Press.

White, R. W. (1959). Motivation reconsidered: The concept of competence. *Psychological Review, 66,* 297–323.

Whiteman, J. L., Young, J. C., & Fisher, M. L. (1985). Teacher burnout and the perception of student behavior. *Education, 105,* 299–305.

Wilcox, B. L., & Vernberg, E. M. (1985). Conceptual and theoretical dilemmas facing social support research. In I. G. Sarason & B. R. Sarason (Eds.), *Social support: Theory, research and applications* (pp. 3–20). Boston: Martinus Nijhoff.

Wills, T. A. (1981). Downward comparison principles in social psychology. *Psychological Bulletin, 90,* 245–271.

Wills, T. A. (1991). Similarity and self-esteem in downward comparison. In J. Suls & T. A. Wills (Eds.), *Social comparison: Contemporary theory and research* (pp. 51–78). Hillsdale, NJ: Erlbaum.

Winnubst, J. A. M., Buunk, B. P., & Marcelissen, F. H. G. (1988). Social support and stress: Perspectives and processes. In S. Fisher & J. Reason (Eds.), *Handbook of life stress, cognition and health* (pp. 511–530). Chichester: John Wiley and Sons.

Winnubst, J. A. M., Marcelissen, F. H. G., & Kleber, R. J. (1982). Effects of social support in the stressor–strain relationship: A Dutch sample. *Social Science & Medicine, 16,* 475–482.

Winnubst, J. A. M., & Van den Bout, J. (1989). Sociale steun en depressie [Social support and depression]. In B. P. Buunk & A. J. Vrugt (Eds.), *Sociale psychologie en psychische klachten* [Social psychology and psychological strain] (pp. 106–113). Assen: Dekker & Van de Vegt.

Wolpin, J. (1986). *Psychological burnout among Canadian teachers: A longitudinal study.* Unpublished doctoral dissertation, York University, Toronto, Canada.

Wood, J. V., Taylor, S. E., & Lichtman, R. R. (1985). Social comparison in adjustment to breast cancer. *Journal of Personality & Social Psychology, 49,* 1169–1183.

Wood, J. V. (1989). Theory and research concerning social comparisons of personal attributes. *Psychological Bulletin, 106,* 231–248.

Wortman, C. B., & Brehm, J. W. (1975). Responses to uncontrollable outcomes: An integration of reactance theory and the learned helplessness model. In L. Berkowitz (Ed.), *Advances in experimental social psychology* (Vol. 8, pp. 277–336). New York: Academic Press.

* Xie, J.-L., & Jamal, M. (1989). *Type A behavior and employee attitudes and behaviors: A study of managers in mainland China.* Paper presented at the annual meeting of Human Relations Management and Organization Behavior, Boston, MA.

Yalom, I. D. (1980). *Existential psychotherapy.* New York: Basic.

Yates, A. J. (1962). *Frustration and conflict.* London: Methuen.

Zedeck, S., Maslach, C., Mosier, K., & Skitka, L. (1988). Affective response to work and quality of family life: Employee and spouse perspectives. *Journal of Social Behavior & Personality, 3,* 135–157.

Zimbardo, P. G. (1970). The human choice: Individuation, reason, and order versus deindividuation, impulse, and chaos. In W. J. Arnold & D. Levine (Eds.), *Nebraska symposium on motivation,* 1969 (pp. 237–307). Lincoln: University of Nebraska Press.

Zuckerman, M. (1979). *Sensation seeking.* Hillsdale, NJ: Erlbaum.

INDEX